Bethlehem Steelmaker

MY 90 YEARS IN LIFE'S LOOP

a memoir by Edmund F. Martin
with David J. Morrison

BMS Press
50 W. North Street
Bethlehem, PA 18018

Bethlehem Steelmaker

MY 90 YEARS IN LIFE'S LOOP

A Memoir

by Edmund F. Martin
with David J. Morrison

Library of Congress Catalogue Card No. 92-082946

ISBN 0-9635251-0-7

First Edition

BMS Press
50 West North Street
Bethlehem, PA 18018

Oaks Printing Company
Printed in the United States of America

Dedicated in loving memory

to my wife

Frances Taylor Martin

Edmund F. Martin

Acknowledgements

This book has been two years in the making, and its successful completion would not have been possible without the help and cooperation of countless friends, colleagues and professionals. Foremost among them has been Barbara Martin Stout, who — as instigator, dealmaker, organizer, fact-checker, editor and coach — piloted the project from idea to reality.

Numerous retired executives and professionals who served Bethlehem Steel with me have added to these recollections during interviews and discussions with the authors. Dozens of hours of tape-recorded interviews, comprising far more material than is included here, will remain available for future historical and scholarly research. Interviewees included James H. Cox (forging and fishing), Lawrence Fenninger Jr. (public affairs), Edward P. and Lala Leach (mining), David T. Dunbar (finance), George A. and Priscilla Payne Hurd (real estate and fishing), James H. McKee Jr.(publications), Lewis W. Foy (chairman), and Millard E. Egbert (chairman's office).

Among Bethlehem Steel's current staff, many were most generous and helpful. Public affairs vice president Stephen G. Donches and corporate communications manager Henry H. Von Spreckelsen provided extensive access to archives and other information sources. Arthur J. Roth, Gary W. Graham, William J. Gignac and Teresa E. Hartnett of the corporate communications office provided friendly and expert assistance whenever asked. Communications coordinator Fanny G. Hartranft provided vast technical assistance and advice. Others at Martin Tower, from the front desk to the 21st floor, exhibited heartfelt pride in their company's heritage by supporting and encouraging the "Martin project" in countless ways.

Erma Buck, my secretary, coordinated appointments and helped with typing and editing. My personal assistant, Debbi Stettler, organized photographs. Attorney Jeffrey A. Parks gave copyright advice, and Ken Oaks of Oaks Printing provided many hours of help and guidance above and beyond his obligation as printer of this book. To others known, unknown or overlooked, we express our gratitude as well.

Contents

Foreword

A lot of people said I ought to write a book. Hell, I'm older than most of you, and I'm older than the Bethlehem Steel Company.

From my childhood experiences on a Kentucky mule farm and amongst the Seminole Indians of Florida to today's global events, I've seen amazing changes and amazing things throughout the world. What an education it's been! Everyone said my father spent half his life in school; Well, I feel like I've spent ninety years in school.

I have a proud, hard-working and close-knit family heritage that makes me the way I am. The Martins and the Fibles were strong pioneer stock. They had good sense about people as well as business. Their story is the story of America. My children, grandchildren and great-grandchildren are continuing that story, each in his or her own way.

My career took me to the top of the most fascinating business in the world. The steel business is about as basic as any business is to our modern society. There are very few things that do not depend on steel in some way. Consequently, the higher I got at Bethlehem Steel the broader my scope of concerns became. All the issues of the day touched the steel business: economics, politics, human relations, the environment, transportation, the infrastructure, education and the fascinating traditions and cultures of the world. As a result, I've had the stimulation of being exposed to new learning experiences every day of my life.

The Loop — Bethlehem Steel's management training program — was established by Eugene Grace in 1922, the year I joined the company. Its premise was to expose newcomers to a rapid-fire succession of learning experiences in every department and phase of the steelmaking business. A little bit of everything.

Its goal was to make us well-rounded. I'd already had quite an education when I joined the Loop — a college graduate at 19 — but I soon realized that you never stop learning.

For 70 of my 90 years I've been affiliated with the Bethlehem Steel Company, its mills, its meeting rooms, its fortunes and its people. However, this memoir does not attempt to serve as a corporate history; historical background is added to illuminate certain issues I encountered, but there are many notable episodes in Bethlehem Steel's history which I have skipped over altogether. Bethlehem's is a history of amazing accomplishments and monumental challenges, and I am proud to have witnessed or taken part in just a few of them.

Those of us whose careers were spent at Bethlehem Steel have much to recollect favorably and proudly. Even with all of the overwhelming and unstoppable adversities that I and my successors fought hard to arrest, and even though it's been many years since the days of never-ending expansion that Mr. Grace took for granted, Bethlehem Steel remains one of the finest organizations that has ever existed.

I wanted the name of Bethlehem Steel to appear somehow in the title of this memoir. When it was suggested that I call myself a "Bethlehem steelmaker," I saw it as a very satisfying honor because it is a designation that I share with tens of thousands of colleagues I have worked with and known over the last 70 years.

I wound up in an office of the company that few people have shared. I was the fourth chairman of Bethlehem Steel. Earlier this year, as my book neared completion, Bethlehem elected its ninth chairman, Hank Barnette. From Charlie Schwab to Hank, I've known each Bethlehem CEO, and I can tell you that all nine of us have been tested, challenged, burdened, stimulated and educated in ways that many people can't imagine.

The subtitle of this memoir, "My 90 years in life's loop," was chosen because the Loop's educational philosophy has been so applicable to every facet of my life: learn something from every encounter and keep moving.

Edmund F. Martin
Bethlehem, PA
November 1, 1992

Chapter One

Kentucky Roots

I grew up as a city boy in Chicago. And I was sick a lot. Every summer they sent me down to the family farm in Kentucky. The whole summer, I wore nothing but blue jeans and a straw hat. Two little colored boys were hired to play with me: Cranston and Karpie. We had a lot of fun together. When I got back home in the fall, the first thing my mother had to do was buy me a new pair of shoes because my feet had grown so much from being barefoot so long.

The Martin homestead, where I spent those summers, was in Bourbon County, Kentucky, about a mile out of Millersburg on the road to Lexington. Near Paris, Kentucky. Paris is a big town. Millersburg I don't think has more than a thousand people now. There were only a couple of stores in the town. Probably still are.

I've been back many times.

Once, after I was married and working for Bethlehem Steel — you'll get a kick out of this — I had to make a quick visit to Millersburg on some family business. I don't remember where we were living at the time — maybe it was Buffalo — but it took me a couple of days to get down there. When I got there, I went to the Martin house in town. My aunt answered the door, and I asked, "Where's Uncle Charlie?"

She said, "He's up at the drug store. Go up and get him."

Bud Wadell's drug store was a popular meeting place. Like

so many small towns in America, the hub of Millersburg's social activity was the drug store.

So I went up and walked in, and I heard some people talking in the back. I thought I'd better find a way to get their attention, so I called out and said: "I want some cigarettes! I'd like a pack of Camels!"

A voice came back and said, "Go buy them somewhere else!"

Uncle Charlie was playing cards with the owner of the drug store, and they weren't about to let a paying customer interrupt their game!

"It's me — Ed," I hollered, "and I've been travelling three days to get here!" Finally, Uncle Charlie came out front, but Bud Wadell never budged!

Uncle Charlie, my father's older brother, was the oldest of Grandfather Samuel Fitch Martin's three sons. It was Uncle Charlie who took over the family farm. He was a very outgoing and colorful individual. When the rest of the family decided Uncle Charlie ought to retire from farming, he moved into town and ran the cemetery. He kept four cows in the cemetery and I don't know how many chickens, so they always had milk at home and chickens when they wanted to eat them! He also became president of the Millersburg Military Institute. When he got to be about 70, one of his sons, my cousin Sam, decided it was time for Uncle Charlie to start taking it easy. He said, "Pop, why don't you come up and spend the winter with us in Boston?" Sam was the buyer for Swift Packing. He could buy a million dollars worth of meat a day if he needed to. Sam's still alive. He's up in New Hampshire. He used to go back to the farm every October, but he hasn't gone for the last couple of years. I think he's a year older than I am.

Anyway, in about a week, all the kids in Sam's neighborhood knew Uncle Charlie. He became a permanent fixture in no time. But, after a few months, he said, "I've got to go home."

Sam said, "What for? It's still the middle of winter!"

Uncle Charlie said, "I've got to sign the diplomas for all the boys at MMI who are graduating in February!"

In those days, the school graduated two classes each year, and Uncle Charlie didn't want even the mid-year class to go out into the world without his signature on their diplomas.

Uncle Charlie finally died at 94. The farm went to his other son, my cousin Edmund. He was named Edmund Hiram, and I was named Edmund Fible. He died in 1990, but his wife Jane still lives there. It's not the original Martin farm, though. That farm was sold when they brought Uncle Charlie in to town to run the cemetery. Then, I think they bought it back again, but they later bought a different farm.

Edmund Hiram Martin was a natural athlete. He had the best coordination of anybody I've ever seen. When we were kids, he could do anything. He would stand up on a horse's back and jump over fences — just standing on the horse's back with the reins in his hand. He was a star athlete at the University of Kentucky. He played everything. He was one of their best football players — not too big — a little bigger than I — but he was good. They used to put three men on him and he still caught passes. He was a good basketball player, too. Nowadays, he would have made a good living as a professional athlete.

Edmund Hiram and I were named after a long line of Kentucky Edmund Martins. It doesn't always come right down from father to son, though, because my father's name was Albert. But he and Uncle Charlie had a third brother who was the Edmund of their generation, so at any given time there were always three Edmunds living. Now, there are just two of us, I think: me, and my cousin Charles down in North Carolina has a son Edmund now.

We're all descended from the first Edmund Martin, who settled in Kentucky after the Revolutionary War.

Edmund Martin the First (1745 - 1811) originally came from New Jersey and was a colonel in the Revolutionary

Army. He met the enemy at the battles of Brandywine and Germantown. Two of his younger brothers served under him, and one of them wrote an account in which he stated that, at Brandywine, "Colonel Martin wore a cocked hat and escaped with his life; for being struck on the forehead with a ball which stunned him, the ball went through the hat, taking part of the flesh."

His hat saved his life! (Almost 200 years later, my hat saved my life when I was thrown through a car windshield in a collision on the German Autobahn. But that's another story.)

Colonel Edmund Martin returned home from the war to find his barns and fences burned and his stock gone, devastated by battle and other wartime activity. An experienced leader, he became a member of the New Jersey Assembly from Sussex County and served as a judge in 1785-86. It wasn't long, however, before he decided that his greatest opportunity lay on the western frontier.

With his wife and ten children, Edmund Martin the First went west. This was in about 1787 or 8. The children ranged in age from Elijah, who was 20, to the second of two Rachels, who was three. Somewhere in Virginia on their way to the Ohio River (West Virginia was not a separate state yet), the party was attacked by Indians. Several of them were killed including Edmund's wife Hannah and a man named Edward Dulin, who was traveling with his wife and child. According to an old account, "this left Susannah Dulin without protection and the ten children of Edmund Martin without a mother. After consultation with other members of the party it was decided that the two should be married when a settlement was reached in which a religious ceremony could be performed."

Apparently, the combined Martin and Dulin households scandalized the entire wagon train all the way to Wheeling, where they finally found a preacher!

On April 29, 1789, at an unknown location, Edmund and Susannah's son, Edmund II was born. This was my great-grandfather, so, had it not been for that fateful Indian massacre, I wouldn't be here today!

Edmund the First eventually settled in Maysville, Kentucky and went into business operating the ferry across the Ohio River. As far as I know, it was quite a successful business. He died in 1811. He's buried in the old Martin family burying ground outside of Millersburg. The graves are all still there. (Most of the later relatives, including my father and mother, are buried in the cemetery that Uncle Charlie ran.)

In 1811, Edmund Martin II, the colonel's son and my great-grandfather, established the family homestead near Millersburg. He was a good businessman and probably the most famous Martin who ever lived in Bourbon County, Kentucky. In addition to the farm, he ran two other successful enterprises. The farm was right on the old turnpike between Maysville and Lexington. It was a good place for a roadside tavern, and the family still has the account books of the tavern he ran from 1812 until he died in 1863. He got all the traffic that travelled the Lexington highway. He also ran a trucking business — with mules and wagons — between Maysville and Lexington. He ran it from the farm there in Millersburg.

They called it a trucking business, because it was not unlike today's trucking industry. It was a big business. There were schedules and record-keeping, teamster payrolls to meet, maintenance of the wagons and the mules, and highway hazards unimaginable today.

Edmund II was a big man. In fact, his size was sort of a Kentucky legend. They tell me he weighed over 400 pounds and was about six feet, eleven inches tall. He could beat his sons running until he was past middle age. They say he used to reach up and put his palm on the ceiling of the tavern and then he'd offer a round drinks on the house if anybody else could do it. I don't think he ever found anybody who could do it, either. It was a good, safe proposition.

The only thing we know about my great-grandmother, Rebecca Stitt Martin (1791 - 1865), is what's written on her tombstone: "She was for 56 years a member of the Methodist Episcopal church. She died as she lived a devout Christian."

She and her 400-pound husband had ten children, the seventh of whom was my grandfather, Samuel Fitch Martin (1828 - 1895). Since his older brothers either died or went west, he inherited the Martin farm, which had increased to 2,000 acres in Bourbon and Nicholas Counties, and the mule business. An old newspaper account says he "was a fine farmer and carried an extensive trade in mules with the Southern planters."

The third Edmund Martin, my uncle Edmund, went into the hardware business in Maysville. Uncle Charlie stayed on the farm and helped run the mule route. My father, who was the youngest of the three, became a lawyer. His father told him: "Albert, you're the only one of my sons who's interested in school work. I set up your brother Edmund in the hardware store, and Charlie's going to get the farm. So I'll pay your way as long as you want to go to school."

My father took full advantage of the offer. I think he went to school until he was 33 years old. He went to Kentucky Wesleyan, where he earned his undergraduate degree, then he went to Cornell for a while. I think he was there four or five years because he liked some history professor. Then, he decided he wanted to be a lawyer, so he ended up at law school at the University of Michigan. He got his law degree when he was 33 years old. He first started practicing law in West Virginia for a year or so, then he went to Chicago.

My parents met each other in Chicago during the 1893 World's Fair, which was also known as the Columbian Exposition, because it opened on the 400th anniversary of Columbus' discovery of America. They met at the Chicago fair, but both my father and my mother were from Kentucky. Their families all lived in Kentucky, and it was "home" to them. We spent a great deal of time there when I was growing up, and all through my life I've gone back to visit every chance I got. After more than 200 years of Martins living there, the roots are pretty deep!

I had the kind of childhood experiences in Kentucky that today seem pretty much like a bygone way of life. The mule route was a big business, and it was still operating when I was growing up. We had a big mule barn right near the highway where they raised and trained the mules. I remember watching Uncle Charlie's teamsters train the mules and teach the teams how to turn. To teach the mules self control, they'd crack a whip next to their ears until they were so accustomed to it they wouldn't flinch. They were the best disciplined mules in Kentucky. Many years, at the state fair in Lexington, the Martin mules won first prize, which was a thousand dollars. They were fine, beautiful-looking mules.

I learned to swim in the mule pond behind the barn. The barn was right next to the highway. Back then, it was just a dirt road, but we called it the highway, because it was the main route from Maysville to Lexington. Extending out the back of the barn maybe eighty feet was this artificial pond. We called it the mule pond because it was next to the barn where the mules lived. It was surrounded by a stone wall, and in the middle there was a rail fence. Out the back of the barn, the water was six or seven feet deep, but up front it was shallow. You could walk right in. You could go along the fence and use it to jump in and climb back out. So it was a good place to learn to swim. But first, I'd have to chase the pigs out! The pigs loved it in there. We kids did, too. I guess I was about six years old the first time I swam in that mule pond.

I never did too much swimming as a kid because, except for that mule pond, we always lived away from water. When we lived in Chicago, my mother took me one time to swim at the South Shore Country Club. The club was right on Lake Michigan, and that's where you swam. Mother was inside the club eating, and I went for a swim. I still wasn't much of a swimmer, but I swam out to the float. When I got up on the float, I realized it would look terrible to crawl down the ladder again. You had to dive in! After all, this was a country club, not a mule pond. So, I'll never forget, the first dive I ever took was at the South Shore Country Club.

At the Martin farm there were old slave quarters still standing (the Civil War had been only 40 years earlier) and out among the slave quarters was a smokehouse. When I got my first 22 rifle, I used to put a target up on the side of the smokehouse to practice my shooting. That's where I learned how to handle a gun, shooting targets on the side of the smokehouse.

There was no such thing as gun control in those days, and it could be argued either way that our learning to handle guns as kids increased the chance of some accidents while decreasing the chance of others. Once, when I shot at a squirrel in a tree, the bullet travelled about half a mile and hit a hired man who was working in the tobacco barn. The bullet took a piece of his pants into his leg, and I think the cloth did more harm than the bullet. We had to get his leg fixed, but everybody seemed to take the accident pretty much in stride.

Actually, it was shooting the targets on the smokehouse that got me into trouble. I didn't realize it, but almost every bullet I shot hit the hams that were hanging inside the smokehouse. When Uncle Charlie came out and caught me doing that, he grabbed my gun and said, "If you haven't got enough sense not to shoot at the smokehouse, you oughtn't to have a gun!"

There were lots of other old buildings I could have used for target practice. I could have put the target on the slave quarters, since there was nothing in them, but I didn't for some reason.

Uncle Charlie took the gun from me and just bent it around so I couldn't shoot anymore. He bent the barrel about half way around! I was about eight years old, but I learned my lesson, and I've been cautious with guns ever since.

We also spent time in the summers at my mother's farm,

which was near La Grange, Kentucky. La Grange is near Louisville, about sixteen miles from Shelbyville. You go through Ballardsville to get there. The farm was called "Brookby," because it was by a brook. Mother bought Brookby after her first husband died — maybe in about 1890 — but before she went to Chicago and met my father.

Mother's maiden name was Fible. The Fibles were from Alsace Lorraine, where the name was originally Fayable. That's what my mother told me. The spelling was Americanized after they moved to Kentucky. Mother grew up in Eminence, Kentucky, where her father owned and operated a hardware store and a department store. Eminence is about midway between La Grange and Shelbyville.

Grandfather Fible also made and sold Bourbon whiskey in Eminence, Kentucky. It was called "Fible & Crabbe," which I always thought sounded like a pretty good name for a Kentucky Bourbon. I never tasted it, though, because by the time I was old enough to drink, Prohibition had started and they'd stopped making it. (I think.)

Both my grandfathers — Grandfather Martin and Grandfather Fible — had good business sense. Each of them ran not one but several successful businesses. They just knew how to run things, how to keep track of everything and everybody. It takes a special knack, and it takes plain, old hard work. Maybe running things is in our blood, because our women are good at running things, too. My mother was always good at running things, and my daughter Barbara is the same way.

I never knew either of my grandfathers. They were both dead when I was born, since there were a lot of years between the generations. The only member of that generation I ever met was my grandmother Fible. I hardly remember her, though. I was still very young when she died. She was struck and killed by a railroad train, but none of us saw it happen.

What a terrible thing that must have been. But sometimes, terrible things make you strong, and I think that's what happened to Mother. It made her strong.

1

2

3

1. Edmund Martin, II, 1789 - 1863. My great-grandfather operated a tavern for 50 years in Bourbon County, Kentucky. He weighed over 400 pounds and was six feet, eleven inches tall.
2. My great-grandmother, Rebecca Stitt Martin, 1781 - 1865. She and her husband had ten children. Her tombstone reads: "She died as she lived a devout Christian."
3. The Martin Homestead is in Millersburg, Kentucky, on the road from Maysville to Lexington. The Martin family mule trucking business was headquartered here.

1. Uncle Charlie Martin. He inherited the family farm. When he retired from farming, he moved into town and ran the Millersburg cemetery. He also served as president of Millersburg Military Institute.
2. Mothers' Day, 1930. My father, Albert Martin, left; my wife, Frances, next to him; and my mother Betty Martin third from right, at a family gathering back in Millersburg. Dad was by this time retired from his position as secretary/treasurer and general counsel for the Elevator Supplies Company in Hoboken, New Jersey.

1

2

1

1. The earliest surviving picture of me. I'm the one in the carriage. At the left are my cousin Sam, who's a year older than me, and a black boy named Johnny, whose family worked on the Martin farm.
2. My parents, Albert and Betty Martin, in Millersburg, 1930.

2

Chapter Two

I Wasn't Always An Umbogas* Poika**

I was five years old when my younger sister Betty was born. One of the earliest things I remember was riding with my mother from La Grange to Shelbyville in a horse and buggy when she was ready to have the baby. My father was away, and I was too little to be left at home, so she just took me with her and drove that buggy all the way to Shelbyville to the hospital!

A few years ago, my daughter Barbara and I went down to Kentucky, and because I'd told her this story so many times, we drove over the same road to measure the distance. It's sixteen miles from La Grange to Shelbyville, which is a long trip by horse and buggy, especially for an expectant mother!

My mother really was a remarkable woman. She was tough and strong-willed, but she was a humanitarian. When I was still only about four years old — it must have been the year before Betty was born — Mother and I lived amongst the Seminole Indians for an entire winter.

My father had bought a farm in Florida where he raised tomatoes and strawberries. It was near Fort Lauderdale on the New River. I don't think he ever went down there. My uncle Robert Bloom (who was married to Mother's sister Sarah) ran the place, and for some reason Mother and I went down and lived there for a while. We lived over the packing shed. It was

(* "good" in Seminole)
(** "boy" in Swedish)

a good-sized building — about a dozen men worked down below — but it was a fruit and vegetable packing shed nonetheless. Our rooms were quite primitive, so we slept under mosquito netting every night. Of course, there was no such thing as spraying insecticides back then. This was around 1906. At that time in Fort Lauderdale, there was only a railroad station, a couple of boarding houses and one store. That's all there was to the town of Fort Lauderdale! Beyond that, there was nothing but farms and Everglades.

Right next to our farm, there was a Seminole Indian encampment. It was so close, we could hear their voices all the time, and we saw them on a regular basis, almost like ordinary neighbors. However, the Indians still lived in their traditional manner, with very few of the white man's customs. It was a pretty tough existence. Every now and then, one of them would come around with a big fish, which was offered in trade for a prepared meal. Without any fuss, Mother usually supplied something. (She was that kind of person. When we lived in Hinsdale, Illinois, a suburb of Chicago, there was a rise in the tracks where the trains always slowed down. A lot of hobos would get off there, and many of them would find their way to our back door. Mother always fed them.)

To a boy my age, living among the Indians was a fascinating adventure. I had a little outfit of Seminole clothes — still have it, somewhere — and I even learned to speak some Seminole. For example, "unga" means "hello," and "umbogas ojus" means "good food."

I remember the Indians used to run on top of the trains at Fort Lauderdale station. They'd climb up and start running over the roofs of the railroad cars just as a train was getting started. Then, when the train got going too fast, they would jump off.

It's hard to imagine it today, but Fort Lauderdale in 1906 was really on the edge of the frontier. I remember that Mother always carried a loaded gun in the horse carriage. She said she carried it because whenever I got out — for example, to open a gate — she didn't want a rattlesnake to bite me. I don't remember Mother ever having to shoot any snakes, but there was never any doubt in my mind that she was ready to do it if she had to!

Mother was fearless; she was prepared for whatever happened, and she faced things head on. Looking back, I can see that she wanted to be sure those qualities were instilled in me. In fact, she may have overdone it!

When we lived in Hinsdale our house was on the edge of town, and there were a bunch of boys from Clarendon Hills, the next town, who used to come by our house and beat me up. I was telling my mother about it one day and she said, "Listen here! When you're here in the house, you're my responsibility. When you're outside, if you get into trouble, you're on your own. I don't want to hear about it!"

So, about a week later, in school I ran into the leader of the Clarendon Hills gang, Billy Wentz, during recess. I started beating him up and I was still pounding him when the recess ended. Because of that I was sent home for a week. I never told my mother about it, and for the whole next week I took my books each morning and went and hid in the woods, pretending to be in school. In the afternoon, when the other kids were dismissed, I'd come down out of the woods and walk home like nothing had ever happened. If Mother ever knew about it, she never told me.

Another time, I got into trouble at Sunday School. We lived about a mile from the Episcopal Church in Hinsdale, so I usually walked to Sunday School with another boy who lived nearby. He and I usually got there before anybody else. One Sunday, I got there ahead of him and put half a pail of water over the door as a practical joke, but instead of him coming through the door, the minister, Reverend Merryman, came through! When Mother got there a short time later, all hell broke loose. I was at that age where mothers usually get totally exasperated by their sons' behavior.

When we got home from Sunday School, Mother didn't say another word, and she didn't give me any dinner either. Later, when my father came home from his Sunday golf game, Mother said, "I can't handle this boy any more. From now on he's your responsibility!" So, after that, I went out with my father whenever he played golf.

That's how I became interested in golf. I started caddying for my father. I remember when his friends used to come for him with a horse and buggy to go to the country club. I was about six years old, but I was always big for my age, so I didn't have any trouble carrying a bag of clubs around the course.

Later, one of the foursome got a car and we would go around to play all the big country clubs around Chicago. It was one of the first cars anybody owned in our area — very much a novelty. We'd roar through all these little towns and farms where they'd probably never seen an automobile before. But the thing that made the biggest impression on me was how many chickens we ran over! I'd always count the number of chickens we killed on each trip.

Ever since those childhood Sundays, I've always been both a dedicated church-goer and a dedicated golfer. I just simply grew up with the idea that that's what Sundays are for! By the time I was ten years old, I knew all the golf courses around Chicago! But I always went to church, too. I did everything in church. I sang in the choir, and later on I was an acolyte. I wasn't old enough to be an acolyte until after we moved to Chicago. We went to the Church of the Redeemer in Chicago, and I remember the minister's name was John Henry Hopkins. We called him "Dodder" Hopkins.

At one point, I was a paid soprano soloist in the church choir. When my voice began to change, I went to my mother and said, "I've got to have singing lessons so I can learn to read music." But she wouldn't let me. She said music lessons were for my sisters. So instead of music lessons I went to the

Art Institute every Saturday for art lessons while my mother went shopping.

As I mentioned before, our house in Hinsdale was on the edge of town, the last house as you went out of town. Beyond that, there were just open fields and the only other thing you came to was a golf club. Now, it's all built up. My father built the house in about 1907, and we lived there for five years or so. It was a large, three-story stucco house with an upstairs porch on the back. I remember that porch very well because once Mother got trapped out there when my sister Betty locked the door.

When we lived there, it was the only house on the block. I think we owned about half the block, so we had room for a big vegetable garden, tended by our Swedish gardener. His name was Sven Strom.

The summer I was about eight years old, that vegetable garden provided my first opportunity for a business venture. I started selling the vegetables, mostly to my mother, but sometimes around town. I also built a hen house and had about a dozen hens. I sold the eggs to my mother, too. Sven Strom used to do most of the work — I was a delegator even then — so for me it was a great learning experience. I even got him to teach me a few words of Swedish. "Svenska poika" is "Swedish boy," and "Svenska flicka" is Swedish girl.

We grew all kinds of things. Most backyard gardeners know that each kind of vegetable has to be grown a certain way. But there are certain ways to successfully sell each crop, too. If I had the first corn of the season, I'd get 25 cents for a dozen ears. If I wasn't the first, I'd only get ten cents. I learned how to grow vegetables from Sven Strom, but I learned all the economic rules on my own.

Every year, I had a crop of pumpkins — big pumpkins, and heavy, like trying to sell steel ingots door-to-door! I learned fast that if I wanted to carry a good selection of pumpkins to my customers, I had to figure out an easy way to do it. Well, there was a family in town that had six boys — the Wells brothers — and they had a donkey. With six boys in the family, I could be pretty sure there was always at least one of them who was looking for a way to get a little bit of spending money. So I'd rent the Wells' donkey from one of the brothers and go around town with a load of pumpkins for sale. I forget how much it cost me to hire the donkey for the day, but I always made money!

I always had spending money, but I never spent any of it. I never really had the desire or the impulse to buy anything. If I earned some money or got a small amount as a gift, I'd get the inevitable question: "What are you going to buy?"

I'd answer, "Nothing."

They'd ask, "Why not?"

I'd say, "Nothing to buy." I guess making money was more interesting to me than spending it.

My parents both had a life-long interest in education. In Hinsdale, they were active in the local school board and they didn't think the schools were good enough. So, before I completed grammar school, we moved into Chicago. I was enrolled in the last class of the University Elementary School, and from there I went on to University High School. Now it's known as the Laboratory School of the University of Chicago.

Many of Chicago's wealthiest industrialists and merchants sent their children to University High, or U High as we all called it.

Julius Rosenwald, the president of Sears Roebuck, sent his daughters to U High, and one of them was in my class. I used to help her with her math homework. Pretty soon, it seemed like all the rich girls in the school were after me to help them with their math assignments!

I think eventually I found myself helping half a dozen girls do their homework. The math professor, Mr. Irvin, would give them 98, but he'd only give me a 95. Of course, I knew my grade was incorrect. One day I caught Irvin making a mistake while he was doing a problem on the blackboard, and I called him on it.

"Martin," he snapped, "you'll never get along in this world!"

We had excellent courses and excellent teachers at U High. There was a prominent scholar in ancient civilization at the University of Chicago. Professor Barnard. One summer, some of us arranged to take a special course in ancient history from him. I think there were three girls and me. One of them drove a big Packard. For a young teenager, I thought that was a pretty great way to spend the summer: going to classes at the University of Chicago with three girls in a big Packard!

U High had wonderful shops. I used to spend my Saturdays and Sundays there every chance I got, and I did all kinds of things. I built a kneehole desk for my mother, which she used for years. When the superintendent of the shops built a motorboat, I helped him, and eventually we launched it on Lake Michigan and motored around Chicago harbor in it.

We had a mechanical shop as well as a woodworking shop. There was an air compressor in the basement. The superintendent was supposed to run it, I guess, but I ran it for him. I used it to make liquid oxygen for the University of Chicago for their chemistry experiments. In my junior year at U High, when I was 14 years old, I bought the castings for a two-cylinder five-horsepower marine engine. I machined it and put it together. While I was boring out one of the cylinders, my finger got caught in a machine gear and was crushed to half its size, so I stuck it in an iodine bottle and it came out all right. Seventy-five years later, I can still see the scar.

The marine engine worked great. A number of years later when my parents had a farm in Southern Maryland, Mother sold my engine to some guy who had an oyster boat. For all I know, it's still plying the waters of Chesapeake Bay.

1

2

1. My sister Betty and me, in front of our house in Hinsdale, about 1910. This may have been the same wheelbarrow I used in my first business venture – selling backyard vegetables to my mother and neighbors.
2. I was a paid soprano soloist at the Episcopal Church in Hinsdale, Illinois. When my voice started to change, I wanted to get singing lessons, but my mother wouldn't let me.
3. My father to the left, our guide, me behind the fish and my uncle and his son make up this family fishing party at Deer Island, Maine. We all spent the summer together there as guests of a client of my father.

3

Chapter Three

College Man

My mother always said to my father, "You're a good lawyer, but you're killing yourself. Here are all these companies offering you a job (as a corporate lawyer); why don't you take one so you don't have to work so damn hard!"

Most of the time he wouldn't get home until midnight.

At the time, my father had a law partnership with Charles A. Williams. The firm of Martin and Williams specialized in municipal legal matters for the city of Chicago, handling things like special assessments and taxation. At one point during this period, my father was one of four special counsels for the city in the important issue of widening Michigan Avenue. The firm also served as counsel to what later became Continental Illinois Bank, one of the nation's biggest banks.

In addition to what surely was very time-consuming government legal work, Dad also lectured at the law schools at Kent State and Northwestern University and was involved in all sorts of civic affairs. He put in long, strenuous hours, and my mother became increasingly worried that he was burning himself out. She felt a corporate legal job would be a lot less exhausting.

Finally, he agreed with her and took a job as secretary-treasurer and general counsel for the Elevator Supplies Company in Hoboken, New Jersey. That's when our family moved to East Orange. It was 1917.

In an account of my father's career which was written after

he died in 1938, there is a comment on his shift to the world of corporate law. It was (in 1917) the era of great industrial expansion, and this led to the development of a type of lawyer who served as special counsel for one concern or at least for a limited number of clients. The example given was Judge Elbert H. Gary, the prominent Chicago lawyer who moved to New York and "engaged his services exclusively to the U.S. Steel Corporation." Indeed, Judge Gary eventually achieved far greater fame and fortune as the head of U.S. Steel — and for that matter the unquestioned leader of the entire U.S. steel industry — than he ever would have otherwise.

I don't believe it was Judge Gary's example that inspired my father — it was my mother's prodding — but it's an interesting possibility. As for me, it was a visit to a steel plant while I was in high school, along with the subsequent advice of a college professor, that attracted me to the steel business.

I was about to begin my senior year at University High in Chicago when we moved to New Jersey. After U High, I was planning to go to Armour Institute, which is now Illinois Tech. However, because of my good grades and the extra credits I earned in summer school, I found that I already had enough credits to get into college. Or so I thought.

With the prospect of moving east, I decided I wanted to apply to MIT or Rensselaer. But I was only fifteen years old, and my mother thought I was "too damn young to go away to college!" She thought I should live at home. So that's how I ended up going to Stevens Institute. It was right near by — in Hoboken — and I could commute from home on the train.

To get into Stevens, I had to take two entrance exams on account of my age. I'd already had four years of high-school mathematics, but Stevens would only give me three years' credit because I had only been three years to high school. So they told me I'd have to take an entrance exam in trigonometry and I think I got a 98 in it. I'd never had any chemistry at U High, because I think we were going to have chemistry in our senior year. When Stevens told me I had to take the chemistry entrance exam too, I arranged to be tutored for about a month by Professor Becker in the Stevens chemistry department. With one month of chemistry under my belt, I passed that exam too.

I was still just fifteen when I entered college. Most of the rest of the boys were coming back from the war, and they were 25 to 30 years old. Some of these guys were twice my age!

I was big for my age, but I was still a skinny kid. I didn't weigh but about 160 pounds. So you can imagine some of the treatment I got from the older fellows when I decided to join a fraternity. I had to grow up fast. I remember one of my fraternity brothers taught me how to shave. His name was Freddy Hale. It was a lot trickier in those days. We still used straight razors.

My fraternity was Sigma Nu. This was in the early days of Prohibition, but we were mostly an athletic fraternity in those days. We didn't go for liquor. What we wanted was steak and milk!

I didn't take a drink until the repeal of Prohibition, and that was more than ten years later. I took a drink then, and I thought it tasted terrible! I couldn't see what all the fuss was about.

Both my parents taught me something I still remember about the danger of alcohol. Their lessons stayed with me. My mother, when I was about 15 or 16, took me down to the basement and she got out a bottle of whiskey from somewhere and held it up in front of my face. She said, "Ed, this is hard whiskey. I want you to be very careful not to abuse it."

And her father was a distiller from Bourbon County.

My father's advice to me also had a lasting impression. Before he died, I remember I decided to ask him: "Dad, remember whenever we would come into the clubhouse from playing golf, and every man had his own bottle of whiskey that the bartender would bring out. Well, you always asked for a glass of buttermilk, and you gave me a ginger ale. Why didn't you let the bartender give you a glass of whiskey?"

"Son," he said, "I learned early in life that I couldn't handle it. I'm one of the lucky ones who learned."

That was right before he died that we talked about it.

I grew up in the right time, I guess. I'm a product of the early days of this century. Life was a lot different. I grew up fast, but I had a good time and a lot of interesting experiences. My college years — age 15 to 19 — were great years.

One of my Sigma Nu brothers was Alexander Calder, who came from the world-famous artist family. He was the one who was known for his suspended sculptures. His father sculpted the statue of William Penn on top of Philadelphia's City Hall.

I played football every year while I was at Stevens. I was a good kicker, because when we had the only house on the block back in Hinsdale, I used to go out there and kick a football around all the time. At Stevens, I was part of the second string who scrimmaged the varsity. The coach would put fourteen or fifteen men on our side to scrimmage the starting lineup. And they were tough, too! Those war veterans weren't afraid of anything!

During the four years I was there, the Stevens football team never lost a game. I think the total score against us was five points, and our total score was 187.

I got involved in a lot of things at Stevens. I was business manager of the yearbook, "The Link," and I was manager of the varsity baseball team one year. I stayed busy, but I had time for fun, too, because that's when I really learned how to organize my time. Time means everything to me.

Every summer during college, I had a summer job of some kind. The first summer — it may have been the summer before I entered Stevens — our family bought a car. I had to build the garage. We had a colored man who helped my mother run the house, so he and I built the garage together. It's a lot easier with two people; you need someone to hold the ladder or help position heavy materials. We made a cement floor, and to this day I can still remember how hard it is to mix all that cement with a shovel!

The next two summers, I got jobs at Elevator Supplies Company. First, I was toolmaker's apprentice, and after that I was a machinist's apprentice, which is a little bit better. Both these jobs were good professional experience, since I was majoring in mechanical engineering.

The fourth summer, I had a memorable business experience. My father got a bid from someone to paint the house. I thought it was two or three times what it ought to be, so I put in a bid too. I said, "I'll paint the house if you pay me half that amount." Well, the summer was almost over and I still wasn't finished painting the house. We were going from brown to white, and I figured maybe it would take two coats, but in many places it took three coats!

So my father said "Why don't you hire somebody to help you get the job done?"

So I hired a friend of mine from college, John Higley, and we got the job done. At any rate, I still made money and I even had time to take a short vacation. I went up to Cape Cod to visit another Stevens friend whose family had a summer place in Brewster. I went up by boat, on the Fall River Line, from New York to Boston, then out to Brewster by train. He and I had a good time there. We drove all over Cape Cod in his father's car. It was a Cole Arrow Eight. I remember it well.

It was a big car. And we hit a cow one time when we were way out on Cape Cod where it was farm country. The farmer had to come out and shoot the cow because it had a broken leg. It damaged the car quite a bit, too, and I remember his father had to get a new car. After that eventful couple of weeks, he and I decided to drive home to New Jersey in his car, which was a Ford with a brass radiator, a Model T, but we just called them Fords, since the Model A wasn't brought out until seven or eight years later. He wanted to start driving home one night after we'd been out until around midnight. So we drove down from Cape Cod starting at midnight in an open car, and that was sort of fun. I think we went through Providence when the sun came up. We finally got home around three or four in the afternoon.

Incidentally, in 1971 after I retired, I drove down to Orange to see that house I painted and the garage I built. The garage is still there, but the wooden walls I built were replaced or covered over, and it didn't look like the same garage at all, but the house was still the same. New paint job, of course.

In June, 1922, I graduated with a bachelor of science degree in mechanical engineering. I graduated in the upper fourth of my class and was elected to Tau Beta Pi, the national engineering fraternity.

Armed with an abundance of energy and a good education from a first-class technical school, I was ready to conquer the world. I was nineteen years old.

1

1. I graduated from Stevens Institute of Technology in 1922 at the age of nineteen. I started my 48-year career at Bethlehem Steel the day after graduation.
2. My father, in 1938, shortly before he died. Someone wrote on the back of this picture, "Taken at Floyd's late in the year. Robert worked the flash and Floyd took the picture."
3. My mother, in 1930. She was a strong and fearless woman. When we lived amongst the Seminole Indians in Fort Lauderdale in 1906, she always carried a gun. It was Mother, not Father, who learned to drive an automobile after I left home.

2

3

1

1. The winning first aid team at Bethlehem Steel. I'm the third from the right. The year was 1923, and I was a draftsman in the roll shop.

2. The program from the Loopers' Picnic on September 6, 1928. I made a speech to the loopers, with Mr. Grace and other executives listening in.

3. My condensed notes for the speech I gave to the 1928 Loop Class.

2

LOOPERS' PICNIC

FOR

MEMBERS AND GUESTS

Seventh Probationary Circuit

OF THE

BETHLEHEM STEEL COMPANY

Picnic at Saucon Valley Country Club

Dinner at Main Office Building

Thursday, September 6, 1928

3

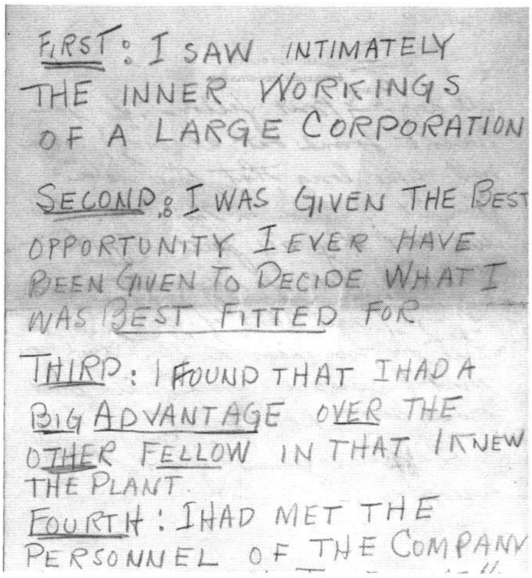

FIRST: I SAW INTIMATELY THE INNER WORKINGS OF A LARGE CORPORATION

SECOND: I WAS GIVEN THE BEST OPPORTUNITY I EVER HAVE BEEN GIVEN TO DECIDE WHAT I WAS BEST FITTED FOR

THIRD: I FOUND THAT I HAD A BIG ADVANTAGE OVER THE OTHER FELLOW IN THAT I KNEW THE PLANT.

FOURTH: I HAD MET THE PERSONNEL OF THE COMPANY

Chapter Four

Starting The Loop

When I graduated from Stevens, I had three job offers. One was at school, teaching physics, and another was with the Otis Elevator Company. (Because of my father's position at Elevator Supplies Company, I knew all the Otis people). The third offer was with Bethlehem. There was a fellow named Charlie Schenk in Bethlehem's order department who had previously been head of the engineering department at Elevator Supplies. He was the one who sent me all the literature about Bethlehem Steel. So I sent a letter of application and I received an offer.

In those days, applying for an out-of-town job was usually done entirely by letter. There was no interview, no plant tour, and no recruiter came to your college campus. Most applicants were considered sight unseen, so a personal recommendation from a third party was very important. Likewise, if we had a choice to make among job offers, we sought out the advice of of somebody we trusted as well. In my case, that's exactly what happened.

I went and asked my chemistry professor, Dr. Pond, which offer I should take. He told me I should take the one at Bethlehem. He said, "If you have a good constitution, go to work for a steel company." So my mother and I drove up here from Orange to look the town over. I think we stayed at the Shawnee Country Club above Easton. I remember we thought Bethlehem looked so nice — not at all like a steel town.

Without hesitation, I decided to come here and take the job.

When I tell this story, some people think it's odd that my mother, not my father, would be involved in formulating my career plans, especially in 1922. It wasn't that my father wasn't interested or that he didn't have a broad understanding of the business world; it was just that my mother had those qualities too. She was a take-charge decision-maker, and her opinion carried a lot of weight in our family. If there were decisions to be made about a steel company, then by golly she was going to help make them. Nowadays, it wouldn't be at all unusual. My mother was a modern woman. She was decades ahead of her contemporaries.

It was around this time that I taught Mother to drive a car. Neither she nor my father had ever learned how to drive. A lot of people of their generation, who had come of age long before there were any automobiles, never learned how to drive. Especially women. As they got older, cars became more commonplace, but they just couldn't learn to drive. Outside of walking or taking trains, if they travelled at all by car they either had chauffeurs, or they waited for other people to drive them around, or they just stayed home. Not my mother! Before I left town and moved away to Pennsylvania, she had me teach her how to drive. She was almost 60 years old.

In 1922, it took more than half a day to get to Bethlehem from Orange. It was almost all dirt roads, and you went right through the center of every town along the way. Sometimes the trip took a full day. I was lucky, therefore, to have had that one visit before deciding to take the job. On my second trip to Bethlehem, I came literally for good!

I graduated from Stevens on June 21, 1922. The very next day, June 22, I came to Pennsylvania and began my career at Bethlehem Steel. I remember it distinctly; I rode the Third Street trolley out to the coke ovens, where I spent my first week as a trainee in the Loop Course. The coke ovens were (and still are) at the far end of the plant, a couple of miles downstream from the Main Plant Entrance. If you didn't have a car, which I didn't, you could get on the trolley to South Bethlehem at the Hotel on Main Street. It would cross the Penny Bridge — the old New Street Bridge — and you'd transfer at Third and New. Then you'd take the Third Street trolley right out to the coke ovens.

That first day in Bethlehem was an eventful one in other respects as well. I came to town with two letters of introduction, as was still the custom in those days. One was to Walter Okeson, a Lehigh University administrator and football referee nationally known as the "czar of football." The other was to Theodore Gray, a Princeton graduate working in the order department at the steel company whose family was from Orange. His father was president of Prudential Life Insurance Company. The letter of introduction was from his cousin Ray, who had played a lot of golf with me at the South Orange Field Club.

It turned out that Theodore Gray was a good person for a newcomer to meet. He knew everybody. He had been such a party boy at Princeton they called him "Rummy Gray," but when he was here in Bethlehem, he never took a drink. That first day in town, Rummy Gray did two things I'll never forget: he said, "I'm taking you out to have lunch at Saucon Valley Country Club," and he introduced me to Frances Taylor, my future wife.

Not bad progress for a nineteen-year-old kid who just 24 hours earlier was listening to a commencement address about life's big challenges and decisions!

There are many stories about my aggressive, impatient, do-it-now personality. Most of them are true, I suppose. However, it is not true that I joined Saucon Valley and played a round of golf that very first day I came to town. I went swimming! And it was more than four years before Frances and I were married.

I did join the club when I was still a "looper." My father used to come up by train to play golf with me on Sundays. He could get to Bethlehem by train in an hour and a half from

Newark. We played pretty nearly every Sunday. The foursome was my father and me, and Theodore Gray and George Fonda. Fonda had been a star football player at Stevens and later made a name for himself by keeping the union out of Weirton Steel after he left Bethlehem.

The only person to question my joining the club while I was still in the Loop was none other than R. A. Lewis, the president of the club and the general manager of the Bethlehem plant. One day, he asked me, "Why would you want to join this country club while you're still a looper? You may not even be stationed here in Bethlehem."

I told him, "I'll take my chances." Then it turned out I had to take a laborer's job to stay here! Charlie Schenk, the fellow most responsible for my coming to Bethlehem Steel, had told me to try to get assigned to the Saucon Division, because that was where all the managers came from.

Back then, when you completed the Loop course, you still faced the challenge of landing a job somewhere within the company. In 1922, we were the first class of loopers. Originated by Bethlehem's president, Eugene Grace, the Loop course would soon earn wide acclaim as the prototype industrial executive training program adopted by many major American corporations. However, my 54 colleagues and I clearly preceded that reputation by several years. In fact, we loopers didn't even enjoy a very good reputation inside the plant! As we rotated through the various divisions and departments of the company, it was clear that we were intruders. Unprepared for our arrival, managers struggled to find ways to keep us occupied. In addition, most of the skilled steelmakers and decision-makers were reluctant to share their talents and procedures with us. I'm sure they were concerned that we would take their jobs or otherwise upset the status quo. As a result, our Loop experience was mostly limited to broom and shovel work.

Nevertheless, I was always proud to have been one of Bethlehem's pioneer loopers. Over my years in the company, I frequently had opportunities to meet with subsequent loop classes. They were always fine young people, and the program quickly became everything that Mr. Grace envisioned. The vast majority of loopers spent their entire career with Bethlehem, becoming solid, devoted leaders in middle and upper management throughout every area of the company. For us newcomers, it was nice to be able to join a big company as part of a group. Regardless of where you ended up, these were your friends, and they stayed your friends. I'm still in touch with Frank Huffman, the only other surviving member of the 1922 Loop class. He's five years older than me. He lives in Upper Michigan now.

The Loop quickly became a much more formal program. I still have the printed program from "the Loopers' Picnic of the Seventh Probationary Circuit" held September 26 at Saucon Valley Country Club. All the top executives of the company attended. The program included golf at 2:45, boxing and wrestling at 3:15 and ball games at 4:00. Dinner was at 7 in the company's main office building, and I was invited to be one of the after-dinner speakers. I still have my original hand-written speech, too!

I believe what I told those fellows was still true when I spoke to subsequent Loop classes as Bethlehem's CEO almost half a century later. To me, there were four main benefits of the Loop.

"First," I told them, "I saw intimately the inner workings of a large corporation. Learning about the various operations from the quarries and mines to the blast furnaces, open-hearths, mills and foundries, and seeing the organization that is necessary to hold a large corporation like this one together. The method of the Loop as it stands today, with talks with the department heads, foremen, and even with men actually doing the work (labor) very nearly forces you to pick up a good knowledge of the steel business.

"Second, the Loop gave me the best opportunity I have ever been given to choose what I was best fitted for. It is hard for nearly all men to pick out what they will work at, especially when they enter college. When a boy doesn't go to college, he usually has had about ten jobs by the time he reaches the age of the college graduate, and knows what he likes to do and what he can do. On the other hand, the college man has had to pick out a particular course in school and he may find that after he tries his civil engineering or mechanical engineering that he doesn't like the drawing board as well as he expected, so he goes into a different line. Now, in going through the Loop, you have an inside view of all these jobs. You see the duties of all, and while you still may pick wrong the first time, you ought to be able to get pretty close.

"Third, after the Loop is over and you get out on a particular job, you fellows will find that you have a big advantage over the other fellow in that you have been through all the plant and will know without any directing where to go to get what you want, no matter what job you decide on. Say, you should go into one of the maintenance departments and are sent out to chase up a casting. You know where the foundry is — also the pattern shop — and after your casting is made you know where to go to get the service department to haul your material where you want it. On the other hand, if you go into the sales department, you know when to find out about special sizes or the time of delivery on a certain job. If you are on the road selling, you have a general plant knowledge that ought to keep you out of many a tight place.

"The fourth advantage I got from the Loop was that I had met the personnel of the company and was able to feel at home. I have worked in plants (summer jobs) where I punched in in the morning, worked all day and went home at night and as far as learning anything about the plant, except the particular job I was on, I might as well have worked in my own back yard all day.

Here, I wrote a note to myself to "PAUSE!"

(I must have worked hard on that speech, knowing that all the top executives, including Mr. Grace, were in the audience. Decades later, I still struggled over the presentation of written speeches.)

"But fellows," I concluded, "if you haven't found out already, you will before long, that there is a fine lot of men in this corporation and that they all pull together. There are lots of arguments and plenty of hell, but if you are here for business — and I suppose most of you are — you will find you will have plenty of help and encouragement from those above you."

I must have tried to memorize that speech, because on the back of the last of my ten small pages I condensed my four points into a few words of large print. I don't remember which version I used, or how it was received. But my message certainly makes sense, even today!

When I completed the Loop in 1922, I took Charlie Schenk's advice. I went up to see Tim Burns who was then superintendent of the Saucon Plant. I was determined to work in the Saucon Division.

The Saucon Division was one of several steel divisions located within the Bethlehem plant. The Lehigh Division, closer to the center of South Bethlehem, was an outgrowth of the original Bethlehem Iron Company, which had been founded in 1861. The Saucon Plant, on the other hand, was an entirely new, Twentieth-Century facility. It was probably the greatest technological achievement of Charlie Schwab's career.

Schwab created the Saucon Plant as the site for Bethlehem Steel's world-famous "Grey Mills." These were mills specially designed to make wide-flanged beams, or I-beams as some people know them. They were anywhere from 18 inches up to 30 inches wide. Later, we made them up to 40 inches wide. The shape enables a much lighter section of steel to carry a heavy load — an ideal design for bridges and tall buildings. In fact, it's often said that the Bethlehem wide-flanged beam "ushered in the skyscraper era." Instead of a fabricated beam, with four angles and three plates, the Bethlehem I-beam was rolled as a single shape, so it was much cheaper, too. The inventor was Henry Grey, an Englishman, but he couldn't interest any steel companies in putting them into production. After Schwab purchased Bethlehem Iron and created Bethlehem Steel, in 1904, he wanted a strategy for building the company to a greatness that rivaled his previous employer and brainchild, U.S. Steel. I think one of the reasons he left U.S. Steel was because they wouldn't build a Grey Mill. They didn't want to risk the money. So that's one of the reasons Schwab came and started Bethlehem, the way I understand it.

Convinced of the potential of Grey's invention, Schwab — in the midst of the Panic of 1907 — gambled his entire net worth, all of Bethlehem's assets and everything he could borrow to build the Grey Mills. His gamble paid off. Schwab and Bethlehem Steel were prosperous beyond all imagination. And the wide-flanged beam fittingly became Bethlehem's corporate logo.

It was in the Saucon Grey Mills that I saw my future.

Amid the heat, the noise and the heavy vibrations that are every steelmaker's daily companions, Tim Burns looked out across the vast expanse of the Saucon Plant, where there were probably 3,000 employees under his supervision. Then he looked at me and said, "I already have a college man. I don't need another one!" He was a typical, old-time master steelmaker of his generation. I had to convince him that I was not a typical, wild-living, pampered college boy of my generation. After a lengthy discussion — I thought it was fifteen minutes but it was probably only about five — Tim Burns capitulated and told me to go down to see George Ackroyd about giving me a job in the Roll Shop. In fact, Burns himself had started in the Roll Shop, and he may have formed the opinion that it was a good place to start a steelmaking career. So that's where I started — as a tool grinder at 30 cents an hour!

An hourly wage of 30 cents meant that I'd be making substantially less than my looper's salary, which had been an even $100 a month. But I didn't have any trouble making ends meet — for example, a caddy only cost 60 cents in 1922 — and I was extremely happy to be stationed in Bethlehem working in the Saucon Division.

As a tool grinder, you had to make the tools so that you could turn the rolls to make the shapes, like beams and angles and so on. You just had to grind the tools so you could make the shapes. Stand at a grinder all day. With goggles on. That was my first job.

A few months later I became a repairman's helper. I was helper to Harvey Rockel. In that job, we repaired the roll lathes. There were tremendous gears that would go around very slowly, just barely moving. Often the studs that held those gears together would pull out. We'd have to bore a bigger hole and put a bigger stud in to hold it.

This was my second job after getting out of the Loop. I'd still only been in Bethlehem a few months. The first place I lived — I can remember it well — was a rented room at 556 N. New Street. My landlady was Mrs. Newhart. I ate my meals at the University Club.

It was still 1922 when I got my third job in the company. The roll shop draftsman was a guy named Spaulding. He was earning $150 a month, and he wanted a raise, but they wouldn't give it to him, so he quit and I got assigned to his job. I saw this as an opportunity to learn a lot, so I worked extra hard. It was very detailed work, actually making little drawings of the precise shape of the rolls that are used to form various sizes of steel shapes — beams, angles, channels and so forth. You had to do it on tracing paper. It was dark, and we worked long hours, so it was really a terrible strain on the eyes. In fact, Ackroyd used to assign me a lot of work that I thought he should have been doing himself. I never knew whether he was trying to teach me how to design rolls or whether he was just plain lazy, but I ended doing all his design work. I remember designing sets of rolls for two new sections, "I-9" and "I-10," that they didn't make up until then. Well I designed those new rolls, and it took a bit of trial and error. They had to throw away the rolls a couple of times until I got them working, but consequently I learned how to design new rolls to make these new sections.

In the meantime, for additional practice, I copied all the designs in the shop, working on my own time, of course. It took me the better part of two years to do that, but I figured it would help me get ahead. I worked 'til eleven or twelve, two and three o'clock in the morning sometimes, copying all the designs, so I could have gotten a job as a roll designer myself — anywhere — after that. Frankly, I had no intention of going anywhere; I just did it because I was always looking ahead.

1

1. A family portrait of the Taylor clan August, 1934. My in-laws, Mr. and Mrs. R. S. Taylor Sr., are in the center. The Martins, Frances and me and Barbara and Bettie, are on the left. On the right are Bob and Miriam Taylor and three of their daughters, Jane, Ann and Mary. R.S. Taylor Sr., a lawyer and banker, was a founder of the Bethlehem Chamber of Commerce and the Rotary Club.
2. Our first house at 449 High Street in historic Bethlehem. It was owned by my father-in-law, who lived across the street at 450 High Street. The R. S. Taylor Jrs. lived next door to them at 442.

2

1

2

1. Beach Haven, 1937. A great weekend of deep sea fishing, beer, sailing and fun. Left to right, Barbara, Charles Herty, Kathleen Herty, Bettie, Millard Dodson, Natalie Dodson, Bob Linderman and me.
2. A tired steelmaker unwinds. This was at Myrtle Beach in October, 1940. I was then serving as assistant superintendent of the Saucon Division under Pat Pazzetti.
3. The Martins at home. Bettie, me, Frances and Barbara at our house at 816 W. Market Street in the early forties.

3

Chapter Five

Saucon Division: Learning to Manage

One day, I went up to see Mr. Cort — Stewart Cort Sr. — who was superintendent of the Saucon Plant. He succeeded Tim Burns after Burns was sent up to run the newly-acquired Lackawanna plant. I'd made up my mind to tell Mr. Cort just what I'd been telling Ackroyd: that if I kept doing that close drawing work another ten years I'd go blind!

I wanted a job in the mills, dealing with people — managing. I'd been asking Ackroyd for a job in the mills, but all he said was, "You can go see Cort if you want, but he won't give you anything."

So I went.

Cort told me: "I'll keep you in mind."

Well, it wasn't too long after that — maybe a couple of months — when one day he caught all the foremen and the clerks and the superintendents of the rollers playing cards in one of the mills. He fired them all on the spot! As a result, I got a job as turn foreman on the twelve- and eighteen-inch mills. It was my first supervisory position, and I had 300 or 400 men working for me. That was in 1924, and I was just 22 years old.

Turn foreman meant I was in charge of one shift. The turn foreman for the other shift was Ike Solliday. He had a big farm down near Hellertown.

At that time, the blast furnaces and the open hearth worked eleven-hour days and thirteen-hour nights. They worked 24 hours a day making steel; from there it came to the mills to be rolled and finished, but the mills could keep up by working just 21 hours a day. So we worked two ten-and-a-half-hour shifts. If I remember correctly, when I was turn foreman in the twelve- and eighteen-inch mills, I'd work one week from 7 A.M. to 5:30, and when I worked nights I started at six. Then, as anybody who worked in those days knows, at the end of two weeks, when you went from the night shift back to the day shift, you worked a double shift around the clock! On those occasions, I'd have to pack a couple of lunches: a loaf of bread, maybe half a dozen porkchops, some hard-boiled eggs, and a quart or two of milk. Anyway, I did that for about a year and a half.

This was long before we had the union, and there was a lot less division between management and labor. When I was working in the mills, Bethlehem Steel had a soccer team, and I used to play soccer with the roll-turners. A lot of them were Europeans and they were good. We'd eat our lunch early, and at noon we played soccer right there at the roll shop in the Saucon Plant.

A fellow that I used to see from time to time, Al Dillman — he was with Keystone Savings & Loan and he was president of the hospital a few years back — his father was one of the best roll-turners we had. I once told him, "I used to play soccer with your father; he was the highest paid roll-turner down at Saucon."

And Al said, "Well at the time he didn't think he made enough!"

In 1926, I was made turn foreman on the 42- and 48-inch mills, the Grey Mills. On the Grey Mills, you had pretty much the same kind of responsibilities as on the 12- and 18-inch mills, only they were much bigger (the two smallest to the two biggest mills) — and much more important to Bethlehem Steel! We made so much money in the Grey Mills — that's where Bethlehem got all the money to buy all those other plants. It was from the profits we made. We couldn't turn out steel fast enough.

Some of the technology in the Grey Mills was German in origin. Consequently, in those days all the people in that mill spoke German. The superintendent of the Grey Mills in 1924 was a bilingual American named Paul Shultz. He was a sailor from Cleveland, and they got him because he could speak German. I could understand his German instructions pretty well, because I'd learned it as a child. One of Mother's sisters had a job teaching crafts in Zurich, and the family spoke German. So, Mother sent me when I was eight or nine years old to a German minister in Chicago to learn German. I can understand it fine, and that turned out to be very helpful in my next promotion at Bethlehem.

Shultz's assistant superintendent was named Harry Costello. When Bethlehem bought the Lackawanna plant, which had been built by the Scranton family up near Buffalo on Lake Erie, the president's office sent Costello up to run the new beam mill that they added. When he left, I got his job, assistant superintendent of the Grey Mills. I was paid $500 a month, which was a lot of money back then. That was in 1927.

As assistant superintendent, I was finally on permanent daytime shift. That's how the organization was set up. Shultz had a daytime assistant — me — and an assistant superintendent for the night shift. That was George Flayhart.

This is a good point for me to tell about my marriage to Frances Taylor. We were married the Saturday after Thanksgiving in 1926. We picked that date for two reasons.

I wasn't aware of it at the time, but there's a story that Frances and I heard many times for years afterwards. I was still on shift work when Frances and I got engaged. Her father, R. S. Taylor Sr., was a prominent lawyer and banker in Bethlehem, and apparently one day he went up to R. A. Lewis, the general manager of the Bethlehem plant, and said:

"When are you going to put that guy Martin on day shift? I'm not going to let him marry my daughter until he's on day shift!"

The reason we got married Thanksgiving weekend was because Costello was an avid football fan. He'd been an all-American at Cornell, and he went to the games every Saturday. Not only did that mean I had to work every Saturday during football season, but it also meant we had to wait until the season ended so my boss could attend the wedding. He wouldn't have let me take a honeymoon either, until the football season was over!

As newlyweds we moved into a house at 449 High Street owned by Mr. Taylor. The Taylors lived across the street at 450 High. I guess they wanted to keep an eye on us. At that point in my career, I appreciated their help. A few years later, I offered to buy the house, but Mr. Taylor wouldn't hear of it. I finally bought it after he died. We lived there until '38 or '39 when we moved to 816 W. Market Street, because we needed a bigger house.

We enjoyed a quiet community life in our early days as a young Bethlehem family. In addition to golf, we were involved in social activities at the country club and in our neighborhood. We played bridge and tennis. At one point, I was asked to be the scoutmaster for a Boy Scout troop at Nativity Episcopal Church. Most of the boys were Catholics from the various South Bethlehem neighborhoods that surrounded the steel plant. I accepted the scoutmaster assignment, but only on the condition that the boys' fathers be permitted to serve on my fathers' committee. (I didn't want any religious or social barriers to hinder their involvement.)

In 1928, Shultz resigned, and I got his job as superintendent of the Grey Mills. That was the year they put me on the bonus system.

When I was in charge of the Grey Mills, a typical day went something like this: first, I'd go down and check on the schedules we were rolling and I would see that enough steel was coming in ingot form, and then I checked on the soaking pits to see that the steel was going to be there in time to get rolled. Then, I had to make sure it was rolled and cut to the right length and then shipped and then ended up in the shipping yard. That was a deal in itself. We'd always have to overroll a bit to make sure we got enough; so then we had a certain amount of stock left over in varying lengths, and you needed to cut that and ship it out when you got additional orders. I think there were 14 runs full of steel down there. It's still operating.

The finished steel all went out by railroad in those days. I forget how many cars a day there were: something like 150 to 200 cars a day. So it was important to have the cooperation of the railroads.

We increased the production to meet the demand while I was there. Up to that time, I don't think anybody pushed it like I did. We had more orders than we could turn out anyway. So we had to do everything we could to increase production just to keep up with the orders. That's the reason we started making a lot of money.

During the Twenties and Thirties, Charlie Schwab was still chairman of Bethlehem Steel, but he gave most of his work to Mr. Grace. He lived in New York mostly, and he didn't come to Bethlehem too often. He used to come to town with his private car on the railroad. And it would be parked down at the South Bethlehem railroad station that was right opposite the Bethlehem Steel office building. He had a big house up near St. Luke's Hospital, but mostly he just stayed in the private car when he was here. It was a magnificent railroad car, which Schwab called the "Loretto" after the town in Western Pennsylvania where he grew up.

There's a story they used to tell in the company dining room about the time Winston Churchill came to visit the Bethlehem plant. When word arrived that Churchill was coming, Mr. Grace called up from somewhere and said, "I can't make it!"

Joe Larkin was the vice president for industrial and public relations, but he wasn't there either. He was playing tennis down at the New Jersey shore somewhere when he got a call from Mr. Grace who said, "Now you get up to Bethlehem right away. Winston Churchill's coming. You give him anything he wants!"

So Joe Larkin came up here, and he still had his tennis pants on and sneakers. And Churchill, the first thing he wanted when he arrived was a drink of scotch. Well, everybody just froze. You couldn't have any drinks in the plant or in any office or anywhere.

Finally, Paul Mackall, the sales manager, somehow produced a bottle of scotch. So they gave Churchill some scotch. And then, while they were entertaining, Joe Larkin said to Churchill, "Mr. Grace couldn't be here, but he told me I should give you anything you wanted. What would you like to have?"

And Churchill said, "I'd like to have Mr. Schwab's private car!" So he took it — after the luncheon — and he went up to Canada and all the way out to the Canadian Pacific and the West Coast and down around San Francisco. I think he had it for three weeks! Charlie Schwab's private railroad car!

Schwab stayed a lot at the big place he had out near his birthplace in Loretto. He also had a big mansion in New York City. When he wasn't at one of those places he was traveling in Europe or somewhere. Schwab didn't spend much time in Bethlehem, but after all, by this time he'd spent forty or fifty years in the steel business and he was enjoying himself.

I remember once when Schwab did visit the plant, I think I was still only a foreman or something, and I wanted to see him. So I'd get up in a crane and look down when he came through. Same way when Walter Chrysler came. It seemed everybody had General Motors cars, but in those days I was a Chrysler driver. When I saw the motorcade of rented Chryslers arrive, I got up on the crane and watched Walter Chrysler go through the plant.

Quite a few years later, I was invited to a dinner party of some kind at Schwab's mansion in Manhattan. It was called "Riverside" and it overlooked the Hudson River. They say it was the biggest private residence ever built in New York City. It cost him millions of dollars to build and a few million more for the furnishings and artwork. This was a large party, of course, and he had some big-name opera singer to perform for us. And there was an organist who played this big pipe organ. Made a hell of a racket! It made every dish and chandelier in the place rattle.

That was Charlie Schwab. He lived like a king.

Schwab apparently spent most of his fortune. He loved to spend money. During the Depression, he kept right on spending, and when people criticized it, his answer was that he was putting money in circulation. What he didn't spend got consumed by bad investments.

The Depression affected the steel business pretty badly. Things really went to hell! In 1930, I was superintendent of the Grey Mills. I remember one month of February when the mill never worked at all, and we had to lay everybody off. There wasn't any business at all.

I remember I'd borrowed some money before the crash to buy some Bethlehem preferred stock. I think I paid $17 a share for it. Then it went down to about $7. But after business came back in '34 I was able to pay off my loan. Eventually, after the war — World War II — the preferred stock went up to $70 or $80!

Under Charlie Schwab's bonus system, managers were paid on production, cost and yield. All three factors went into determining how much of a bonus you got. Even though business slowed down in 1930, I earned more in '30 than I did in '29. In fact, I made way more than I was supposed to make. I think it was $10,000. So they cut my rate, and I got mad, and I went home and stayed home a week!

Business had slowed down, but the way my bonus was figured, I beat it! One thing I remember was, we used to have copper bearings for the side mills on the beams. So I'd save all this copper until I had, oh, I don't know how much, but tons of it. I'd save it until the price of copper went up, and when it went up I'd sell it. That's one of the things I did to beat the bonus system!

There was another time too when they cut my rate. It was years later, when I was general manager up at Lackawanna. Someone showed the executive office that I made more than I was supposed to. So they cut my rate that time, too!

The bonus system was a good thing, and it motivated us to produce and save money. But it was always an unwieldy system that kept getting out of control. Years later, when I became chairman, we finally had to get it under control once and for all.

At one point during the Depression, things were so bad we just didn't get any bonus at all. I think my pay went down to $80 a month. Down from $800! We managed somehow. Some people were a lot worse off than we were. I remember there was one guy in the mill who was so broke, with six kids to support. I gave him 20 of my 80 bucks each month so he could stay alive.

In 1939, they promoted me to assistant superintendent of the Saucon Division. Pat Pazzetti, who had been superintendent of the open hearths, was my boss. (Pazzetti was another boss of mine who had been an all-American football player. He's one of Lehigh's all-time football heroes.) Because the Saucon Division was almost the equivalent of an entire steel plant, being superintendent of the whole division was like being a plant general manager, which was the title of the top man at Lackawanna, Sparrows Point and our other plants.

The Saucon Division not only had the Grey Mills for making structural shapes, but we also had ten open hearth furnaces and two tilting furnaces to make all the structural steel. And for the rails — we also made rails — there was a Bessemer to make steel for rails.

I remember the labor strike of 1941. I was not yet in a policy-making role at Bethlehem Steel, so I won't go into the theoretical issues. It was a very big and very tense, but very short (four days) confrontation, which is usually the scenario when a union is getting organized or established at a particular plant or industry. I was assistant superintendent down at Saucon, and I was asked to find a place to stable the State Police horses for the mounted troops that had been sent in. I also had to arrange to buy hay to feed the horses. That job was delegated to me.

But I got along fine with the workers and the strikers. I knew most of them, and I'd worked with them. One of the fellows who scheduled the orders coming into the rolling mills was head of the Employees' Representation Plan, which was the company union that was the forerunner of the United Steelworkers. I had to talk with him every day about the orders that were coming in, but we never talked about his union business. He did that on the side. That was in the time period before the strike. I guess he helped organize it.

The result of the strike was the establishment of the United Steelworkers union at Bethlehem. And I got along fine with them after that.

Bethlehem Steel was never bigger than during World War II. I think we had something like 350,000 employees during the war years. But a lot of them were at the shipyards. We produced steel for the war effort in record-breaking amounts, and the shipyards built and repaired thousands of ships. Our steel tonnage increased, even above the wartime records, during the Fifties and Sixties, but we never had as many employees. That was because of the wartime shipyard activity.

Bethlehem Steel really helped win the war. We no doubt shortened it. I was assistant superintendent of the Saucon Plant until about a year after the war ended.

1

1. Not-so-junior steel executive. This was taken in 1951, around the time I was made general manager of Bethlehem's Lackawanna Plant. I was 49 years old, and I'd been with the company 30 years.
2. Our house at 227 Middlesex Road in Buffalo, where we lived from 1947 to 1957. When we were transferred to Lackawanna in 1946, the post-World-War-II housing shortage forced us to spend six months in the Statler Hotel. We then lived briefly at 8 Arlington Place.

2

1

1 and 2. We really enjoyed our years in Buffalo. I bought a boat, the "Skimmer" and joined two boating clubs on Lake Erie. In these pictures, taken in July, 1954, Frances and I are seen taking Barbara and our grandson, Peter Stout, for a spin.

2

1

2

1. Ned Entwistle, my boss when I arrived at Lackawanna, had been sent up there from the Saucon Division about ten years ahead of me. He and his wife, Fern, helped make Frances and me feel at home in Buffalo.
2. University of Buffalo, June, 1961. Newly promoted to the presidency of Bethlehem Steel, I was invited back to Buffalo to receive an honorary degree. In addition to my success with the Lackawanna Plant, I'd gotten involved in a great many civic affairs in Buffalo.
3. Bethlehem Steel's Lackawanna Plant at Buffalo, New York, where I served as General Manager from 1949 to 1957. We tripled steel output to 6 million tons a year and made the plant profitable.

3

Chapter Six

Lackawanna: Challenge and Change

In 1946, they sent me up to the Lackawanna plant as one of two assistant general managers. The eleven years I spent there were as challenging as any other phase of my steelmaking career. The plant was considered something of a lemon. It lost money every year, and nobody who had been sent up there had been able to turn it around. Consequently, getting sent to Lackawanna was a lot like being traded to a losing baseball team. It was a real career-killer.

I had hoped that when they sent me away, I would go to Sparrows Point, but, sure enough, they sent me up there to Lackawanna!

I guess I'd reached that stage of my career with Bethlehem when management felt it was time for a decisive test of my abilities, and of course they also wanted very desperately to turn Lackawanna around. Well, it was a test, all right, but Ed Martin and Lackawanna both came out on top. From losing money, we ended up by making a hell of a lot of money! The Lackawanna plant became the third largest steel plant in America while I was there. We tripled our steel output, from two million tons to six million tons a year.

Lackawanna was one of the steel plants Bethlehem purchased with its profits from the First World War and the Grey Mills. Charlie Schwab had a vision of a vast steelmaking organization that would include every sector of the business —

ordnance, structural steel, rails, sheet steel and everything else. It would be an organization that had every possible geographic advantage, including access to ocean shipping and the Great Lakes, and it would own all its raw material sources. The mastermind who made that dream a reality was Eugene Grace, who was elevated to the presidency of Bethlehem Steel Corporation by Charlie Schwab in 1916. Lackawanna fit into their strategy on all counts.

The plant at Lackawanna has an interesting history. Lackawanna is an Indian name meaning "fork in the river." In the vicinity of Scranton, Pennsylvania, in Lackawanna County, there is a fork in the Lackawanna River. It was at this location that a small enterprise was established in 1840 for producing nails, mostly for mineshaft construction in the local anthracite coal industry. The company was beset with technical and financial problems until it was rescued a few years later by a group of investors that included J. C. Platt and the Scranton brothers, George and Seldon.

Even with the infusion of new investment capital, the Scrantons' business remained unsteady for several years until a boom in railroad construction offered new opportunities. In 1847, with an order for 12,000 tons of rails for the expanding New York and Erie Railroad, the firm expanded its rolling mill and reorganized as the Lackawanna Iron & Coal Co. Despite the inferiority of iron to steel for making rails, Lackawanna iron rails soon gained a reputation for their durability. With the steady expansion of railroads during the mid-century and Civil War periods, the company prospered. It was at this time that Moses Taylor began a long and successful career with Lackawanna.

Iron rails were never entirely satisfactory. Despite the superior design of Lackawanna iron rails, they were cursed by the inferiority of local raw materials. As a result, Lackawanna was among the first firms to switch to making steel rails.

Although England's Henry Bessemer is largely credited with the discovery of modern steelmaking, the same technique was simultaneously perfected in 1857 in Johnstown,

Pennsylvania, by William Kelly. In fact, many historians claim that the Kelly Converter actually superseded its better-known British counterpart as the world's first technique for converting pig iron to steel. The new technology was perfectly timed to have a monumental impact on the development of the world's railroads.

The Civil War delayed the adaptation of the new steelmaking technique at American plants. In the interim, American railroads desiring the longer-lasting steel rails had to import them from Britain. This was unacceptable for a lot of reasons, not the least of which was the close relationship between American railroads and iron companies. In those days, there was a railroad in every ironmaking family tree, and vice-versa.

America's ironmakers began installing steelmaking facilities at the close of the Civil War. In 1866, America's first steel plant was established by the Pennsylvania Railroad near Harrisburg, where their main line connected with the Susquehanna River. They named the enterprise the Pennsylvania Steel Company, and they named the town "Steelton." Pennsylvania Steel soon merged with Maryland Steel, whose main plant was at Sparrows Point on the Chesapeake Bay. Ultimately the combined enterprise was bought out by Bethlehem in 1916 as part of Mr. Grace's expansion program.

Responding to the demand for steel rails, the Lackawanna Iron & Coal Co. installed its first Bessemer furnace in 1875. This became a subsidiary, the Scranton Steel Co., in 1881. In 1891, the firms were reorganized as the Lackawanna Iron and Steel Company. Soon the company was producing 500,000 tons of steel a year. Before the turn of the century, however, the owners launched plans to relocate the business closer to premium raw materials and primary transportation systems. One place they considered was Lebanon, Pennsylvania, where the Lackawanna firm owned an interest in steel furnaces and the historic Cornwall ore mines.

The location which offered all the advantages they were

seeking was up on the southern shore of Lake Erie near Buffalo. It had access to major railroads as well as lake shipping, and it enjoyed convenient proximity to midwest markets and the superior-quality iron ore of the Western Great Lakes region.

In 1900, the year of the great Pan American exposition at Buffalo, work was begun just a few miles away on a project that attracted its own share of attention as one of the country's biggest construction undertakings on one site. One of the first and certainly the biggest plant since Steelton to be built from the ground up, it was of significant interest to steel works engineers from throughout the country. The January 7, 1904, issue of *Iron Age* praised the 1400-acre waterfront complex as "the world's greatest individual steel plant." Others called it the Eighth Wonder of the World.

The company was reorganized with a capitalization of $60 million, and construction continued through the first full decade of the Twentieth Century. Soon it was a complete and fully integrated steel mill. It boasted numerous innovations including the use of electricity for many functions previously done by steam or hydraulic power. The company controlled extensive coal and ore mines in Pennsylvania, ore holdings in the Mesabi range in Minnesota and a long list of smaller iron and steel production facilities. With titans like Cornelius Vanderbilt seated on the Board along with the Scrantons and Moses Taylor, Lackawanna Iron and Steel was one of the strongest industrial enterprises in the country.

During World War I, Lackawanna produced four and a half million tons of steel, most of it destined for American armament. During that same period, the company reported earnings of $114,000,000. Over the years, however, the Lackawanna owners engaged in a program of compulsive expansion which consumed most of the company's earnings. Much of this reinvestment took place without regard for technological advancements occurring elsewhere in the steel industry. For all its modern features and Twentieth-Century innovations, Lackawanna quickly lost whatever competitive edge it might have had. In less than two decades, it had become an outdated money-loser, barely surviving the big 1919 steel and coal strikes.

There was talk of selling Lackawanna Steel or including it in a proposed merger of six independent steel firms (Midvale, Youngstown Sheet and Tube, Republic, Inland, Brier Hill, and Steel and Tube of America), an idea that many thought would create a worthy competitor for U.S. Steel.

What happened came as something of a surprise. Fifteen minutes before the stock market closed on May 11, 1922, Eugene Grace issued an announcement that Bethlehem Steel was purchasing Lackawanna, a move that would unite under a single management the two largest independent steel producers in the United States. ("Independent" meant anybody other than U.S. Steel.) Lackawanna's assets on December 31, 1921, had been listed at roughly $90 million. Mr. Grace paid $60 million.

Lackawanna's investors were at least glad to get their principal back. Bethlehem's management was delighted at the idea of becoming a major controller of Great Lakes iron ore and steel production. Strategic as the move was, Lackawanna soon proved to be an impossible stepchild whose potential remained just beyond the reach of Bethlehem's parental control.

Lackawanna produced a lot of steel — during peacetime the principal product was rails — but it was a perennial money-loser because the managers Bethlehem inherited from the previous owners simply weren't running the place right. They didn't have the Bethlehem system. And they had terrible superintendents throughout the plant!

A major part of the Lackawanna complex was the sheetmaking operation of the former Seneca Iron and Steel Company which Bethlehem had acquired in 1932. Seneca had been right down the road from Lackawanna, so Bethlehem just moved the whole Seneca sheetmaking mill right into the Lackawanna plant. This was the first plant that Bethlehem owned that made wide sheets, which are used in making automobiles, household appliances, office furniture and things like that. Bethlehem hadn't made any sheets until they bought Seneca. So they got a fellow — his name was Shonkweiler — who they hired from one of the other sheet companies, and they installed him at Lackawanna as assistant general manager. That was just before I arrived as the other assistant general manager in 1946. Although I had been with Bethlehem a lot longer than he had, Shonkweiler had seniority over me in terms of our current positions. Over both of us was the guy who was responsible for the entire Lackawanna operation, Ned Entwistle, the general manager. His wife's name was Fern, who Frances and I got to know through company social occasions during our few overlapping years in Buffalo.

Ned Entwistle had been my superior as superintendent of the Saucon Division. I think he came from the Cambria plant at Johnstown before that. I guess he was sent up to Lackawanna in about 1936 — ten years ahead of me. Shonkweiler and I were brought in, not only in anticipation of Entwistle's retirement, but our arrival at Lackawanna coincided with the company-wide shuffling and reorganization that was launched after World War II ended.

That reorganization began right at the top. In December, 1945, Mr. Grace was elected Chairman, filling a vacancy that had existed (at least on paper) since Charlie Schwab's death in 1939. I guess because it was wartime, and out of respect for the company's founder and only chairman, Mr. Grace no doubt felt it would be awkward to make that change any sooner.

At the same time of the announcement that Mr. Grace would become chairman (and continue as chief executive officer), it was also announced that Art Homer would succeed him as president. He'd been responsible for Bethlehem's incredible shipbuilding accomplishments during the War, fulfilling Mr. Grace's promise to build a ship a day for Uncle Sam! S.J. Cort, who had been down at Sparrows Point for a number of years as general manager, returned to Bethlehem and became vice president of steelmaking operations. He replaced Quincy Bent when Bent retired.

Anyway, back at Lackawanna, when Ned Entwistle left in 1947, they picked Shonkweiler to be the next general manager. They expected great things from him, primarily because of his background in sheetmaking. He was supposed to develop our sheet business. Shonkweiler really knew the business, too, but he had a serious personal problem: he was an alcoholic.

After a couple of years they sent him to Florida on some nebulous assignment. I think he was there about a year when Cort told me, "We're not going to bring Shonkweiler back. We're going to make you Acting General Manager." But he didn't tell anybody right away. So I had the responsibility for running the plant and supposedly the authority to make decisions, but only a handful of top executives back in Bethlehem knew it. Which is a very awkward way to have to run a steel plant that's in dire need of a lot of major changes!

Finally Cort announced to everybody that Shonkweiler wasn't coming back and I was going to be the new boss. That meant I could begin to make the changes that Cort and Mr. Grace expected. We were then able to move away from Shonkweiler's concentration on sheetmaking. One of my first major projects was to help put in a wide-flange beam mill.

In 1946, Mr. Grace announced his peacetime expansion plans, a companywide, long-range modernization and expansion program with a projected cost of $134,000,000. (He ultimately spent about three times that amount.) This included improvements at every location Bethlehem owned, most of which had already just undergone significant expansion to meet wartime needs. At Lackawanna, for example, wartime expansion had included two new batteries of coke ovens and two new blast furnaces.

In announcing his companywide improvement plans, Mr. Grace emphasized that Bethlehem's goal was to increase efficiency as well as capacity in order to meet worldwide postwar competition. Bethlehem's management was positioning the company not only to help rebuild the free world over the short run but also to compete in it over the long run. The improvements we made at Lackawanna were certainly tied to that strategy. Between 1946 and 1950, we just about rebuilt the plant from the ground up.

Among other improvements, we installed a continuous hot scarfer between the blooming mill and the billet mill. Like a planing blade in a sawmill, the scarfer shaves off a thin layer of the surface of the bloom in order to produce a higher quality product. Next, the billet mill which feeds the bar mill was rebuilt with alternately horizontal and vertical rolls in order to eliminate the turning or twisting that is otherwise needed for shaping all four sides of the bloom.

A big project was the building of an entire new 10-inch bar mill adjacent to our 12-inch mill. The new mill replaced Lackawanna's old 8- and 10-inch mills and incorporated the same feature to avoid twist as at the billet mill. We rebuilt and enlarged the soaking pits, and we modernized and upgraded the capacity of the strip-sheet mill.

In time, the Lackawanna plant became one of the world's largest producers of cold-rolled sheets, producing more than a million tons a year of high-grade cold-rolled steel sheet for such growing post-war industries as automobiles, appliances, galvanized products and modern architectural "curtain wall" building exteriors.

We installed a wide-flange beam mill, just like the Grey Mill back in the Saucon Division in Bethlehem. This gave the company a major share of the structural steel market in the Great Lakes region. The Grey Mill quickly became a success, and soon we were making as much money on structural steel as we were making on sheets. Before long, the Lackawanna Plant became as profitable as any plant Bethlehem owned.

When I first got to Lackawanna we were having a strike a week. Sometimes more. The worst problem was in the coke ovens. It was constant chaos in the coke ovens. Nothing went right. And all the bricklayers were unionized, too, and some of the strikes came from them.

The head of the union up there was a full-blooded American Indian named Julian Bruce. He and I got along very well. I went to his house many times. He'd be eating dinner and his wife would stand up in back of him. It was just their custom that she wasn't allowed to sit down while he was eating. She'd just stand there serving him, and nobody said anything about it because that was the way the Indians did it. I liked Julian Bruce, and he and I got along fine.

One of the difficulties we had at Lackawanna was a severe labor shortage. The Lackawanna plant was about three or four miles outside of Buffalo, which was the main source of labor. But there were two other big steel plants there, Republic Steel and — I forget the other one. Anyway, they competed with us for men.

It's hard to imagine today that we ever had a problem like that in the steel industry, other than in wartime. This was in the postwar period, and the American steel industry was still reaching its peak. At Lackawanna, we tried to build our workforce up to 22,000 employees while I was there, but the most I think we ever got was 20,000. We even got the state to build new highways out from Buffalo — bridges and everything — so we could attract the workers we needed, but we never got the number we wanted to have. As a result, we had terrible overtime expenses.

I finally had to fire the head of the coke ovens. His name was Bob Pugh. In fact, I almost got fired too! I told Pugh, "The union's running the coke ovens and you're not!"

Finally Cort came up and said, "Ed didn't know anything about coke ovens and he had to learn, but he knows now." I had to bring the Koppers people up to straighten it out and I learned from them. They built coke ovens all over the country. In fact, they originally owned the coke ovens at Bethlehem. Bethlehem only bought them after they had them going.

After I found out I'd have to fire Bob Pugh, I promoted his assistant, a guy named Meyer. I said to him, "Listen. Forget all you ever knew about coke ovens, and do 'em altogether different than you have been. Just forget everything!"

Meyer was good. He ended up becoming general manager of the whole Lackawanna Plant!

The problem in the coke ovens was the fires. The oven doors leaked, and the air movement caused fires to flare up. You can't have that kind of fire in coke ovens. You want the coal to bake in a very controlled manner, so you've got to stop the leaks that cause the fires. I'd say to Meyer, "You've got to keep these fires out of the doors." I was always telling him that. I used to drive over to the coke ovens three or four times a day after I found out how bad they were. Morning, noon and night, I'd drive around the coke ovens looking for any fires I could see. If I saw any fires, I'd call Meyer and he'd get them put out. Meyer did a pretty good job solving that problem, and his assistant, McCord, was a good man, too. I still get a Christmas Card from him every year!

The Lackawanna plant had its own coke ovens from the very beginning, but they had to be rebuilt several times. I think when I rebuilt them, it cost the company something like five or six million bucks, and that's part of why I almost got fired.

Cort almost fired me another time, too. That was when I was elected president of the Wanakah Country Club in Buffalo. He didn't think it was right for me to do that. He had another man — I think it was Granger from Sparrows Point — that he was grooming for my job. It was between me and Granger, he said. But I stayed president of the club anyway, and I wasn't fired.

After Cort didn't fire me, he saw it didn't make any difference. In fact, I'm still an honorary member of the club!

What's even more surprising is that I didn't get fired when I built the Bethlehem Steel Club up at Lackawanna. Some people thought I made it too nice. Those were the days after the war when business was booming everywhere and we were making lots of money and our people were all prosperous. Bethlehem Steel built country clubs at every plant, so I built the one at Lackawanna. I bought the land; we had a good location right on Lake Erie. The course cost a million dollars, and I think the clubhouse cost a million and a half. The best architect in Buffalo designed the building. For the course I used the same designer, Gordon from Quakertown, who designed the Grace Course at Saucon Valley a few years earlier. I remember walking all the holes with him when we laid out the course. The only hole I can still remember in detail is the 17th; it was 185 yards and had traps on both sides.

A few years ago Bethlehem sold the club when they were selling off real estate and other assets. It's now privately owned, and they call it the Briarwood Country Club. In the summer of 1992, a startling discovery was made. The people up at Briarwood somehow discovered that the guy who built their club was still alive, and I was surprised to discover that they were about to have a golf tournament named for me: The 31st annual Ed Martin Tournament! It must have been started a couple of years after I moved back to Bethlehem. They wanted me to come up and present the trophies. I couldn't make the trip, so we videotaped a little speech in my office in Bethlehem, and they showed it at the awards banquet.

Our years in Buffalo were especially happy years for our family, despite the upheaval of moving. At first, the thought of moving from Bethlehem was greeted with distaste. Our daughter Barbara was about to begin her senior year at Liberty High School; she was involved in all sorts of activities, and she hated to leave.

Frances, of course, loved Bethlehem, her native city. We had just bought the house of her dreams: at 24 E. Market Street owned by an old maid named Annie Gauff. It was a very old house in the historic part of town. It had been built and owned by the Moravians before it was owned individually. She'd had her eye on it for years. It was full of antique furniture, and we bought the house with pretty near all the antiques. Frances was still decorating when she found out we had to leave. The day I told her we had to move was the only time she said she wished she hadn't married a steel man.

Then, to make matters worse, when we got to Buffalo, we couldn't find a house to buy. It was right after the war ended, and you just couldn't find houses. There were none for sale. So we ended up living six months in the Statler Hotel in downtown Buffalo. With the whole family living there for six months, and eating our meals there and all, I think the bill ended up being $100,000. It was such a big bill, I was afraid for the longest time to submit it to the company.

The bill for our furniture storage was $800 for six months. I took it off my income tax, and the I.R.S. questioned the deduction. They asked me, "Why didn't your company pay for this?"

I said, "After they paid to put up the whole family in a hotel for six months, I didn't have enough nerve to ask them to pay for the furniture, too!"

That was the only time I've ever had any kind of trouble with the I.R.S.

Finally we bought a house at 8 Arlington Place. On the day we moved our furniture in, the man across the street came over and invited us to have lunch with him. He was Ralph Hindman, the Presbyterian minister at the First Presbyterian Church of Buffalo. He became one of our dearest friends. We even joined his church. I was an Episcopalian and Frances was a Moravian, but we went to Ralph Hindman's Presbyterian church. And I even became a trustee of the church.

We all got involved in things in Buffalo. Frances joined a number of clubs and charities. She was president of United Cerebral Palsy, and Bettie worked there as a receptionist. I became a director of the Chamber of Commerce, the Better Business Bureau, the Erie County Planning Association and Buffalo Port committee. It was fun living on Lake Erie, too. I hadn't had much opportunity for boating for quite a few years, so we took advantage of our proximity to the lake. I bought a runabout, the "Skimmer," and joined the Buffalo Launch Club and the North Shore Yacht Club up in Ridgeway, Ontario.

We ended up loving Buffalo, and the one who loved it most of all was our daughter Bettie. Dr. Hindman got her involved in the young people's fellowship, and she really had the time of her life.

1

1. Bigger fish. After I returned to Bethlehem and became operating vice president, we bought a house in Fort Lauderdale. Like most steelmakers, I'd always believed in working hard and playing hard. I especially enjoyed the rare opportunities I had for deep sea fishing, like this outing with the renowned fishing guide, Jimmy Albright.
2. Another fish and another fishing buddy, Carr Parker.

2

1

1. The house we bought when we moved back to Bethlehem. It was right across the street from Mr. Grace's house on Prospect Avenue, which was known as "Bonus Hill" because of its popularity with steel executives. Both the Graces' house and ours are now owned by the Catholic Diocese.
2. Charlie Schwab founded Bethlehem Steel in 1905 and served as chairman until his death in 1939. He lived like a king and owned the largest private residence in New York City.
3. The legendary Eugene Grace who joined Bethlehem after graduation from Lehigh in 1899. He proved himself in the iron ore mines of Cuba, became president in 1916, and succeeded Charlie Schwab as chairman in 1945. He said he and I were "the only working men" who won the golf championship at Saucon Valley.

2

3

1

1. Heirs to the Grace legacy. This was the foursome that ran Bethlehem Steel during Eugene Grace's declining years. Norborne Berkeley was vice president of administration; I was operating vice president; Bob McMath was vice president of finance; and Art Homer was president. Berkeley and McMath were long-time Grace confidantes who never really accepted Homer as their boss. It was an awkward time.
2. President and Chairman. Art Homer compiled an outstanding record as head of Bethlehem shipbuilding, but as CEO he relied on my steelmaking background. It wasn't easy following in Grace's footsteps, but Homer's greatest success was in helping to modernize the corporation.

2

Chapter Seven

Back To Bethlehem

My return to Bethlehem's home office occurred in 1957, a milestone year in the history of Bethlehem Steel Corporation. It was the year Eugene G. Grace withdrew from active management of the company after more than forty years as its chief executive officer. It was also a year of record-breaking revenues ($2.6 billion) and record-breaking earnings ($191 million).

Mr. Grace's life's work had earned worldwide recognition and respect both for himself and for Bethlehem Steel. His life-long philosophy — for himself, for Bethlehem and for the industry — had been "onward and upward." Every limit was a challenge to be overcome, and every peak or record was simply a new goal to be surpassed. Whatever you charted in Mr. Grace's world, the curves only went one way: up! The thought of perpetuating (not to mention surpassing) that greatness posed a nearly impossible challenge among the fellows of my generation. Very soon, we'd have that challenge and more.

On December 20, 1957, Bethlehem Steel announced the retirement of two vice presidents, S. J. Cort and H. C. Crawford. Both were long-time members of Mr. Grace's inner circle and outstanding steel executives in their own right.

The announcement, which was made by president A. B. Homer, further stated that their successors would be E. F. Martin and F. M. Huffman, respectively.

(Bethlehem Steel had a strict habit of referring to their people by their first initials. It was a mandatory form of address in correspondence, press releases, company phone books — everything. This was a fairly common way of showing respect in those days, but we carried it to extremes. In the company phone books from the early Twenties, I'm just plain Ed Martin. But when I reached a certain level a few years later, they began identifying me as E. F. I think they always made an exception for Quincy Bent, though. He didn't have a middle initial, so, rather than listing him as Q. Bent, they always spelled out his name.)

H. C. — Harry — Crawford was probably the top traffic expert in the entire steel industry. He'd spent his whole career in traffic work, initially at other steel companies, including Carnegie, Cambria and Midvale. When Bethlehem Steel acquired the Cambria properties from Midvale in 1923, Harry Crawford was brought to Bethlehem as assistant general traffic manager.

"Traffic" of course refers to the endless network of railroads, rail yards, rail cars, and contents of rail cars that traditionally served the steel industry. You couldn't simply assume that all those millions of tons of raw materials and finished steel products would end up where you wanted them and at the hour they were expected. Just watching a long freight train, with its variety of contents all destined for different places and its array of cars all displaying different corporate logos, it's easy to imagine the almost unsolvable puzzles. Of course the railroads also have traffic managers, but their priorities and problems never coincide entirely with those of the customer. As a result, heavy users like steel companies had their own traffic departments to monitor and second-guess the whole process. It was quite a job, and it included marine shipping as well. We'd buy ore by the boat, bringing it up from Chile or from Africa or somewhere. There would be maybe 550 carloads after the boat was unloaded; 500 would come to Bethlehem, and we had an office full of clerks trying to find out what happened to the other fifty cars. And they could be anywhere; some of them would even end up on the West Coast! In Harry Crawford's day, I think he had twelve men just to track down lost cars.

In addition to managing Bethlehem's global shipping responsibilities, the traffic department established its own in-house railroads to move things around on plant property. We owned and operated seven different railroads, each one working on the premises of a different steel plant. The names always amused people; for example, the Philadelphia, Bethlehem & New England Railroad served the Bethlehem plant, but of course it never went anywhere near Philadelphia or New England.

Traffic was at its most challenging during Harry Crawford's career — before the days of computers! Harry played a key role in helping to orchestrate the crucial movement of steel during World War II. He was elected vice president in charge of traffic in 1947, retiring ten years later. Under the men who succeeded him, Frank Huffman (who'd been in the Loop with me), Tom Church and others, the business has changed a great deal. Today, Bethlehem still relies a lot on rail shipping, but trucking has received an increasing share of the work. Trucks are a lot simpler to manage. They go where you send them! And they can get almost anywhere in the country overnight. Other changes, notably in information technology, have continually improved traffic management — and the expectations placed on it.

Reflecting the elimination of those old-fashioned traffic snarls (or at least most of them), the traffic department was redesignated as the transportation department in 1971.

S. J. Cort had been a boss of mine for much of my career at Bethlehem. I succeeded him as vice president in charge of steel plant operations upon his retirement, effective January 1, 1958. I was honored to be following in his footsteps. When it came to steelmaking, few people were as talented as S. J. Cort. His contributions to the modernization of the steel industry received national recognition, including the first Benjamin F.

Fairless Award of the American Institute of Mining, Metallurgical and Petroleum Engineers. He no doubt would have advanced to the presidency or chairmanship of Bethlehem Steel if it hadn't been for the logjam of old-timers who stayed on in the executive suite during Mr. Grace's last years.

It wasn't only S. J. Cort's steelmaking talents that kept people like me in awe as we followed him through the ranks. He had the strictest code of personal behavior of anyone at Bethlehem Steel, and he expected the same standards in his subordinates. Unlike Mr. Grace, Cort was more than just a teetotaler, he was practically a temperance crusader! I think this reached its peak in the late Twenties or early Thirties — the last years of Prohibition — when Cort was general manager at Sparrows Point. Of course, those were the days when a plant general manager had almost as much control over your personal life as he did at work. Anyway, Carl Hoffman (the fellow who had been my roommate at the University Club in our early days with the company) was working at Sparrows Point as superintendent of the blast furnaces. And when my daughter Bettie was in the hospital in Baltimore, I used to go down there to see her and I'd stop off and visit Hoffman. If we were going to have a drink, he'd have us sit on the living-room floor, because Cort used to walk around the town and look in windows to see who was drinking. And often he'd fire them!

My appointment to succeed Mr. Cort came at an interesting time in the history of Bethlehem's senior management. Just a few months earlier, at the April 1957 quarterly board of directors meeting, the board was enlarged from 15 to 21 members with the election of six new directors. All of the appointees were Bethlehem Steel men, consistent with Mr. Grace's long-standing policy of composing the board entirely of inside directors. The six were: Russell Branscom, assistant vice president, industrial and public relations; Bernard Broeker, secretary of the corporation; Paul Entrekin, assistant vice president, mining; Francis Huffman, assistant vice president, traffic; William Johnstone, vice president, real estate; and Ivor Sims, assistant vice president, purchasing. These men were senior staffers by almost any measurement — most were over 55 — but they were relative youngsters in the boardroom in 1957.

Many of the other fifteen directors had served for decades, starting with Mr. Grace himself, who had joined the board as a vice president in 1911! Robert McMath, Norborne Berkeley and Joe Larkin had all been on the board since the early Thirties; Art Homer had been a director since '40 and president since '45; and there was a group that included James Bender, Daniel Strohmeier, Donald Aikenhead, Arthur Peterson and Paul Killian, as well as Harry Crawford and S. J. Cort, who all had joined the board in the late Forties.

Most of these men had never worked for anybody but Eugene Grace. After all those years under one boss, they would very soon be taking part in the historic transition to Art Homer's chairmanship in 1960 and then to my own four years later. Sims would continue even further into the administration of Stewart S. Cort (the son of S. J.); and Broeker's service on the board would span 20 years and five chairmanships, including that of Lewis W. Foy.

But we're getting ahead of the story, because back in the spring of 1957 I was neither a board member nor vice president, and all these men were still my superiors!

I remember it was a very awkward time. It was right around the time when all these other promotions were announced that Art Homer told me that I was going to be brought back from Lackawanna and made vice president for steel plant operations.

Homer said, "We're going to bring you back to Bethlehem and make you operating vice president. Now I want you to go around and find out — at all the other plants — how they're operating so you know what you're doing when you come down here!"

Among all these senior-level changes, I'd been picked for

the most important vice presidency at Bethlehem Steel, but they didn't tell anybody but me! So I had to go around to all these other plants, and people wondered why I was there, but I didn't tell anyone. I went to all the steel plants and shipyards — I think fifteen places in all. I spent four or five months seeing the entire Bethlehem Steel Corporation. It was sort of a private, do-it-yourself loop course, and it was every bit as unrehearsed and impromptu as the first one I'd been on 35 years earlier.

Finally, in December my appointment as vice president was announced. A few weeks later, on January 30, 1958, I was elected to the board of directors, filling the seat that had just been vacated by S. J. Cort. The *Bethlehem Globe-Times* stories carrying these announcements were among the first steel company articles in almost half a century not to contain any attribution or mention of Mr. Grace. During the summer of 1957, he'd had the first of a series of strokes which began to curtail his abilities.

In those days, Bethlehem didn't have any mandatory retirement age. Schwab had remained chairman until he died at age 78 or 79. That was not unusual back then. There were lots of companies where the top guy never stepped down, and Mr. Grace was one of the last of that breed.

When I joined the board of directors, Mr. Grace's vitality had faded, but he still chaired some of the meetings. After all, he was over 80 and weakened by one or two strokes. Sometimes he would nod off to sleep, and everybody wouldn't dare do anything until he woke up. Sometimes an hour would go by. We didn't leave. We didn't continue with whatever business was being discussed. We just waited in silence until he woke up and resumed the meeting as if nothing had happened.

Everyone had far too much respect for Mr. Grace and his accomplishments to do any different. He had been at Bethlehem Steel close to 60 years. Actually it was more than

60 years when he died.

Grace was a tense and emotional man. He could get upset very easily. He was a smart man, though, and we all admired him for that. He had great power of concentration, he was a hard worker, and he was proud of his background as a "working man," which was the term he used for someone who worked his way up through the steelmaking departments. One of his first jobs at the steel company was operating a crane. In 1937, when I won the club championship in golf at Saucon Valley Country Club, I remember Mr. Grace said to me: "Martin, you know, you and I are the only two working men who won this." It was a greater accomplishment in his opinion, because our careers hadn't allowed us as much time for golf as other people got.

(I think the only reason I played golf so well in '37 was the fact that Frances and my daughters were in Beach Haven for the summer; on weeknights, I didn't want to go home to an empty house, so most evenings I headed for the golf course instead. In '37, I played Harry Maxwell, the superintendent of the roll shop, on the final 36 holes. I beat him on the last hole when he drove into the creek on Number Eighteen.)

On another occasion, Mr. Grace came up to me at a party and admired an honorary society key that I was wearing. "I've got one of those, too," he said. There are few higher compliments than when a great man makes a point of acknowledging some attribute that you have in common with him.

I had less in common with Art Homer, because he was a shipbuilding man rather than a steelmaker. He'd spent the early part of his career at Bethlehem's shipyards at Sparrows Point and Quincy. At the beginning of World War II, Mr. Grace put him in charge of all of Bethlehem Steel's shipbuilding operations. It was the largest shipbuilding enterprise in the country, if not the world. At the height of the war, Grace made the memorable announcement that Bethlehem was going to deliver new ships to Uncle Sam at the rate of one a day, a commitment that would stand for the duration of the war. It was Art Homer who made good on that commitment.

Homer had always been interested in ships. He'd been in the submarine service during the First World War, joined Bethlehem Steel in 1919 (just three years ahead of me), and spent most of his career in the shipbuilding division. He became vice president of the shipbuilding division in 1940.

When the War ended, Homer's reward was the presidency of Bethlehem Steel Corporation. He was only 49 years old. In December of 1945, Mr. Grace became Chairman, finally filling a five-year vacancy that had existed since Charles Schwab's death in 1939.

When Grace became chairman, he naturally continued to run the company, establishing a pattern whereby the day-to-day management of the company would be the chairman's responsibility. Schwab, as founder, had served as chairman for 35 years. Although he skillfully orchestrated the company's national and international affairs, he never maintained regular office hours. He could be gone months at a time. He had vast amounts of figures and data committed to memory, so he could remain in complete command of the company wherever he happened to be, without any need for paperwork. They say he could calculate all kinds of cost and production figures in his head.

Schwab and his spontaneous management style of course were long gone when I became part of Bethlehem Steel's senior management. Mr. Grace was pretty much on the sidelines in failing health, and Cort had just retired. These steelmaking experts — who had built Bethlehem Steel into one of the biggest industrial concerns in the country — were gone, and all of a sudden I had become the senior steel man in the company. It was a big leap in responsibility. Considering how slow the changes had been in the executive suite up to that point, I could scarcely imagine that I'd be president in two years and CEO less than four years after that.

As much as we'd enjoyed Buffalo, Frances was delighted to

be returning to Bethlehem. We bought the house on Prospect Avenue right across the street from Mr. Grace's house. It had been Ted Bickford's house. Then Bethlehem Steel bought it, so we bought it from them.

A lot of the houses of Bethlehem's executives stayed in the company. They were passed down from one senior officer to another, with the company sometimes acting as middleman. When I was president or chairman, they offered me Grace's house, but it was just too big. I think Lehigh ended up inheriting it from Grace, but they sold it and now it's a nursing home. And the nuns who run it live in our old house across the street. We sold it to the Catholic Diocese. I'd put in a lot of fancy stereo equipment; I left it all in the house and donated it to the church. I went back a number of years later and they were still listening to it. They kept the place up pretty well, but they never could keep the hedges trimmed as evenly as I did.

They used to call Prospect Avenue "Bonus Hill," since so many steel executives lived there. But some of them always lived out in Saucon Valley, too. We moved out there in '63 or '64. We bought Sylvester's house, "Oak Knoll," right opposite the entrance to the country club on Saucon Valley Road. I paid $125,000 for it.

After Homer moved away, Bethlehem Steel bought his house and used it for a few years to put up outside directors and other visitors when they were in town. We used it until we got Weyhill from Mrs. Bent. When Homer had it, I remember he had a special soundproof room for private conversations. The Foys own Homer's house now.

Quincy Bent's farm became the Weyhill Country Club. It was 225 acres, and he sold it to the company with the provision that he and his wife could live there for the rest of their lives. I put in the golf course when I was chairman. It has 18 holes. They're completely separate from Saucon. Many people think Saucon is owned by Bethlehem Steel, but it's not. It never was.

1

1. Making headlines. After 40 years in steel plants dealing mostly with company men, I was suddenly thrust in the national spotlight after my famous, "Hell, no" remark at the 1962 stockholders meeting. A lot of people still insist I never said it.
2. Art Homer, who fulfilled Mr. Grace's commitment to build "a ship a day" for Uncle Sam during World War II, was president from 1945 to 1960. He succeeded Grace as chairman and retired in 1964 in accordance with his new retirement policy.

2

Chapter Eight

Making Headlines

One of the things that happens when you become CEO of a company like Bethlehem Steel is you're all of a sudden a public figure. You say something, and it makes headlines. Eugene Grace was a national figure, and Charlie Schwab was actually an international figure. His trips to Europe were always front-page news.

Schwab and Grace each had a public image that was many years in the making, and that's part of the reason they're both still so well known today. Most modern chief executives simply aren't in office long enough to achieve that kind of widespread recognition.

My first memorable experience in the national spotlight was completely unplanned and unexpected. And it was well before I became CEO. It happened in April, 1962, when Art Homer was still chairman and I was in my second year as president.

That year, I had to preside at the annual Bethlehem Steel stockholders' meeting down in Wilmington, because Homer was home sick in bed. The meeting itself went fine, but afterwards, a remark I made to a reporter made headlines all over the country.

A few days earlier, the steel companies and the union had signed a new labor agreement providing wage increases over the next two years, so the reporter asked if me that meant prices would go up.

I said, "Hell no! This isn't the time to raise prices."

Back then, it was almost automatic that all the steel companies would raise prices after each wage increase — every two years or so. After World War II, the demand for American steel was so strong that it accommodated a steady spiral of wage and price increases. The same thing was happening in the automotive and other major industries. The Fifties was the decade when the American dream was being pursued and achieved across the board. After fifteen years of depression and war, suddenly the constraints were no longer in place, and a lot of people felt they had a lot of catching up to do. This of course is the surest recipe for inflation.

The Kennedy Administration wanted to curb inflation, and they felt the steel industry was the place to start. President Kennedy's secretary of labor was Arthur Goldberg, who is best known to the American public for having subsequently served as ambassador to the United Nations and as a justice of the Supreme Court. But, to us in the steel industry, he was better known as the former chief attorney for the United Steelworkers of America. He was probably the best labor lawyer of the 20th Century (who had a lot to do with winning all those wage increases from us.) Now he was President Kennedy's point man for resolving the contract dispute between the steel industry and the union. When your old opponent comes back as the referee, look out!

Over the years, Goldberg had done his job for the Steelworkers exceedingly well — perhaps too well. Between 1940 and 1958, employment costs in the steel industry increased at the rate of about 8% per year, significantly outpacing inflation. By 1958, American steelworkers were virtually the highest paid industrial workers in the world, earning an average base pay of $3.03 per hour. While that seems low today (nearly a quarter-century later), it was far above steelworkers' wages in Japan and Western Europe. The average Japanese steelworker's wage was 42 cents per hour — less than one-seventh the American rate. All European steelworkers were averaging less than a dollar per hour. American steelworkers' wages also exceeded those in other American industries by a substantial margin. In addition, steelworker benefits — pensions, insurance, Social Security, vacations, shift premiums and supplementary unemployment benefits — were always way ahead of other industries.

Through the Fifties, the demand for American steel, both at home and abroad, was so strong that customers were willing to pay any price. In those days, we couldn't make steel fast enough! Our customers were so concerned about making sure they got a steady supply of steel, they'd send their buyers to stay at the Hotel Bethlehem for months on end just to make sure they got their orders filled. The hotel was always full of people trying to get their steel.

The bubble burst in 1958, when we were hit by the severest recession since the Great Depression. Heavy industry was hit especially hard. At Bethlehem Steel, production plummeted to about 58% of capacity. Our total output dropped from more than 19 million net tons of steel in 1957 (a record production year) to about 13.3 million net tons in 1958. Between inflation, recession and the growing threat of foreign competition, the steel industry faced tougher times unless it took corrective action.

An opportunity for corrective action presented itself in early 1959 with the opening of negotiations for a new labor agreement. As an anti-inflationary initiative, the steel companies proposed a no-increase contract. It of course was rejected. Then we proposed giving modest increases in exchange for the elimination of certain wasteful practices. That too was rejected, and on July 14, 1959, an industry-wide strike began.

That strike, by interrupting the supply of steel and by creating lingering doubts among our customers about the future availability of steel, caused many steel customers to establish links to overseas sources of steel. Up to this point, what little foreign steel was coming to America was viewed as inferior product. The price was not the issue. But when the strike of

1959 forced them to shop overseas for steel, they found that most of it was as good as American steel, and it was cheaper. That was really the beginning of the import problem.

After two or three more years of inflation in the early Sixties, the handwriting, at least in my mind, was clearly on the wall. American steel prices were increasing much too quickly, and foreign producers were securing a distinct foothold in the American steel market. The old pattern of the Fifties had burned itself out, and the trouble was starting for the American steel industry.

The Kennedy Administration blamed the inflationary spiral on the steel companies, and we blamed it on, among other people, Kennedy's top labor man. Mix in the usual politics, and in hindsight it's easy see that a clash of some kind was inevitable.

Ed Martin unwittingly provided an opportunity for that clash to occur by making an off-hand remark to a bunch of reporters at the conclusion of the 1962 Bethlehem Steel stockholders' meeting.

My remark made headlines across the country. Because raising prices after signing a new labor contract had become the familiar pattern, the media assumed I'd let the cat out of the bag about a change in policy.

I wasn't referring to a policy change. I was simply expressing an unrehearsed opinion off the top of my head. In my mind, it was more important than ever that we keep steel prices down. With the first truly modest union wage increase since the War, and with competition steadily increasing from overseas as well as from non-steel alternative materials, my opinion was that it would not make sense to raise the price of steel. But things really heated up when U.S. Steel announced that they were raising their prices, and Bethlehem matched the increase, contradicting what I had said just one day earlier.

That night, I had my memorable F.B.I. wake-up call. At

ten o'clock at night, three F.B.I. agents came to our house on Prospect Avenue. We later heard they had been sent by Robert Kennedy, the Attorney General, who was as scrappy as Goldberg. The F.B.I. guys knocked, and my wife answered the door. Their gameplan was to catch me at an unguarded moment. They figured I might be groggy and therefore easier to question, and they figured that at home I wouldn't have the usual phalanx of outer-office people running interference. I guess they also figured that such a startling visit would scare me into being cooperative.

The only person running interference for me was my wife. Frances was a very small woman, and I guess they thought they could bulldoze their way right in. But dealing with her was a tougher challenge than they anticipated!

The F.B.I. agents asked to see Edmund Martin, and Frances responded: "Mr. Martin's office hours are eight to five; if you want to see him on business, you can see him tomorrow at his office!"

I forget exactly what happened next, but I think I ended up talking to them for about half an hour. I was in my pajamas, but I didn't care. They were very polite. They wanted to know exactly what I said at the stockholders meeting, so I repeated: "Hell, no, this isn't the time to raise prices." I think they figured there was a lot more to it than that, but there wasn't.

When Bethlehem joined U.S. Steel in raising prices on certain products the very next day, the news media zeroed in once again on my remark. In response, the official company statement was that "Mr. Martin was, in fact, indefinite about the matter of prices."

The "Hell, no" remark continued to have reverberations, though. On account of the fuss made by the White House — in which the Kennedy Administration sought to establish itself as the arbiter of steel prices — there was a big backlash on Wall Street. Steel stock prices were depressed for a few months, and the brokerage people all lashed out at Kennedy for what they saw as an anti-capitalist posture.

After 30 years, there are stockbrokers who still discuss the incident when my name is mentioned. What a way to become a public figure!

Some people think I got upset or embarrassed over that incident, but I didn't mind. I was telling the truth. When you tell the truth, usually you come out all right.

Incidentally, I never had any direct dealings or encounters with President Kennedy. The only time I saw JFK was at a football game in Miami — the Orange Bowl, I guess. We used to get invited by somebody on the committee, so I always had great seats. One year, I was on the fourth or fifth row and Kennedy sat almost right in front of me. I remember he lit a great big cigar and smoked it. He seemed too young to be president.

I always thought maybe Homer skipped that stockholders' meeting on purpose, in order to see what kind of a job I'd do running a stockholders' meeting. He hadn't yet made it official that I would succeed him as chairman, so perhaps this was a way to test me. That's how I viewed it, anyway. Homer was thinking of Bethlehem Steel's future.

Art Homer deserves a lot of credit for thinking about Bethlehem Steel's future. He wasn't a steel man, and he never claimed to be. But he was a good executive. His greatest accomplishments as chairman were future-oriented.

Arthur Homer served Bethlehem Steel for 44 years. He was best known for his outstanding shipbuilding record, which reached its zenith during World War II. However, following that successful achievement in running the largest shipbuilding operation the world has ever known, Homer spent the next phase of his career in a status that was quite awkward. Some people said he was in a "twelve-year loop course," observing the management of Bethlehem Steel more so than participating in it.

Homer had spent almost all of his career away from the Bethlehem office, at various shipyards and at the New York office at 25 Broadway. Consequently, when he moved into the sixth-floor executive suite in 1945 as Bethlehem's third president — but without Schwab's or Grace's world-class steelmaking credentials — Homer was viewed as an outsider.

For the next twelve years, Grace continued to run the company as he always had. He still had most of his old vice presidents, who were long accustomed to dealing directly with Mr. Grace, and vice versa. They'd become a tight-knit little group, in and out of the office. It wasn't easy to change their routine at this stage of the game, so Homer was simply bypassed much of the time. Not only was he on the periphery of the steelmaking side of things, but Homer also spent these post-war years watching the shipbuilding side steadily shrinking in importance. I was at Lackawanna throughout this period, but Homer's difficult situation was no secret. It's possible that Grace even planned it this way on purpose so that he wouldn't have to share his throne with anyone. Certainly Grace's vice presidents favored this arrangement; it meant they would continue to have access to Mr. Grace. In addition, they avoided competing amongst themselves for a promotion that would have upset a long-standing balance of power.

Even during the last three years of Mr. Grace's life, when he was incapacitated by ill health, the torch was never completely passed to Art Homer. Grace held the title of "honorary chairman" until his death on July 25, 1960.

Eugene Grace's long affiliation with Bethlehem Steel had given him the unchallenged and awe-inspiring aura of royalty which no one dared diminish as long as there was a breath left in him. His funeral in Packer Chapel at Lehigh University (where he also was viewed as royalty, having chaired the university's board of trustees for over 40 years) was like that of a head of state. All the board members of both Lehigh University and Bethlehem Steel (of which I was one) served as honorary pallbearers. From across the nation, prominent personalities from industry, government and higher education filled the enormous gothic chapel. There was extensive coverage by the news media, including solemn-voiced radio reporters. At the hour of the funeral, some 150,000 Bethlehem Steel employees in every steel plant, shipyard and office across the country observed a two-minute period of (relative) silence.

Grace was buried in Bethlehem's Nisky Hill Cemetery on a prominent slope which has a commanding vista of the sprawling steel plant and home offices spread along the far bank of the Lehigh River. Bill Johnstone, the vice president who had helped to facilitate many of Mr. Grace's real estate and construction projects, was responsible for yet another: the elegant, amphitheater-like, final resting place of the man who had been Bethlehem Steel's center-stage legend for more than half a century.

Within a few days of Mr. Grace's funeral, Art Homer was elected chairman of Bethlehem Steel Corporation, and the changing of the guard at long last took place. At the same directors' meeting — August 1, 1960 — I was elected president. John E. "Jake" Jacobs succeeded me as operating vice president. Jake Jacobs also had succeeded me in 1958 as general manager at Lackawanna.

Homer had begun to make changes during the three years of Mr. Grace's incapacitation. That was a sticky time for Art Homer, too, because there was a public impression perpetuated that Mr. Grace was still very much in control, although this was not the case. Homer was definitely the chief executive, the buck stopped at his desk, and he ultimately was responsible to the stockholders. In addition to assuming this awesome responsibility, he had the added burden of exercising it without appearing disrespectful of the ailing Mr. Grace. Under these circumstances, he did remarkably well.

Nobody likes to tell people they must retire, but that's what Homer had to tell quite a few people. Having endured twelve years in executive limbo with the silent, expressionless poker-face for which he was known, Homer surely saw Bethlehem Steel's lack of a retirement policy as more harmful to the company than to him personally. Here he was — chief executive officer and in his early sixties — and he had a whole array of vice presidents and other men surrounding him who were ten and fifteen years older! On almost a case-by-case basis, Homer had the unpleasant task of informing these old-timers that the time had come for them to retire. He started with the men who were 75 and older, and in a few years' time he had it down to 70.

In 1961, Homer announced the retirement of the three men who had been Mr. Grace's top assistants for nearly thirty years, Robert E. McMath, Norborne Berkeley and Joseph M. Larkin. They were superb executives who had helped Bethlehem Steel rise to greatness through depression, war, prosperity and expansion, but they were clearly part of an outdated regime. The actual end of the Grace regime occurred on December 31, 1961, the date Norborne Berkeley retired. He had been Grace's administrative vice president, alter ego, and a director since 1935. He was a small man, like Grace, and he had been a star midget quarterback at the University of Virginia. Berkeley, rather than Homer, had occupied the big office connecting to Mr. Grace's. As Grace's top assistant and closest confidante, Berkeley was a crucial component of Grace's executive machinery. Whatever Mr. Grace wanted, it was Berkley who got it, one way or another. Much of Grace's power flowed through Norborne Berkeley.

Homer's ultimate objective was the adoption of a permanent, companywide mandatory retirement policy. So, at the July, 1963, board meeting we made 68 the mandatory retirement age for directors and 65 the retirement age for everyone else.

Four more directors were affected immediately by the new policy, and they retired during the next few months. Interestingly, that opened up spots on the board for the two men who eventually would follow Homer and me as chairman of the corporation: Stewart S. Cort and Lewis W. Foy. Cort also was named to succeed me as president, and Foy succeeded Paul Killian as vice president for purchasing. The other two board slots went to Edward Bickford, who succeeded Jess Honeycutt as vice president of sales, and Albert Reed, who succeeded Frank Brugler as comptroller.

S. S. Cort had spent almost all of his Bethlehem career on the West Coast. Over the years, so much of our Pacific-area activities were consolidated under his jurisdiction that he became, in 1961, the first and only man to have the title of vice president for Pacific Coast Operations. He held that post until he came back east in August, 1963.

As part of this whole package of inter-related chessboard moves that resulted from the adoption of the new retirement policy, Cort was named president effective August 1, 1963; I was made vice chairman; and Art Homer's retirement was scheduled to occur the following April, when he himself would reach the now-mandatory retirement age of 68. Although our subsequent promotions were by no means a foregone conclusion (a chairman still has to be elected into office by the other directors), Bethlehem Steel now had, for the first time in its history, a relatively smooth and predictable line of succession. The credit for making this long-overdue correction must go to Arthur Homer.

The irony of the situation was that Homer didn't want to retire any more than anyone else did! He'd been held in limbo for so many years that he was already almost 65 when he finally became chairman. He spent less than four years in the job. But he wanted the rules to stick, so he didn't try to make an exception for himself. As it turned out, I was the last person to retire at 68 — Bill Johnstone was next to last — before we further lowered the retirement age for directors to 65. I stayed right up until my 69th birthday! Heck, we all loved working at Bethlehem Steel. I'd probably still be there if it weren't for Art Homer's retirement rules!

Another important contribution that Art Homer made to

Bethlehem Steel was the importance he placed on our research program. Again, he was thinking about the future. Having been at the center of the monumental transition from the Eugene Grace era, he was uniquely sensitive to the company's needs as it faced an increasingly complex future.

Schwab and Grace had managed by instinct, self-confidence and sheer force of personality. Both of them were brilliant thinkers and planners, who culled information and opinions from their top men, but took full responsibility for settling upon a final course of action; first Schwab, then Grace charted the course of Bethlehem's growth and prosperity with the solitary and supreme status of a ship captain. Homer knew that neither he, nor anyone else for that matter, could continue that solitary method of calling the shots.

And shots were getting more difficult to call. As costs, risks, competition and consequences increased, decisions increasingly would have to be founded on painstaking scientific analysis and in-depth research. That could only be done if research efforts were placed into a sophisticated organization structured in such a way as to have meaningful interaction and influence with the operating and decision-making side of the company. A research organization like that needs first-class tools, resources and facilities; it needs first-class space to think and work effectively; and, to assure its influence, it needs clout and turf and prestige.

In short, a first-class research program needs a first-class headquarters, and that's just what Art Homer decided we should build. The Homer Research Laboratories at the summit of South Mountain stand as a well-deserved monument to Bethlehem Steel's third chairman and chief executive officer.

The idea of establishing a research center on the mountaintop was first explored in 1957 in a study undertaken in collaboration with Lehigh University. Lehigh's relationship with Bethlehem Steel dated back many decades. Eugene Grace had been in Lehigh's class of 1899, and we've been hiring Lehigh graduates ever since. A lot of the research done by Lehigh's science and engineering people over the years has had close ties to Bethlehem Steel. Because of these ties, it was decided that everybody would benefit if we put the research center on top of South Mountain adjoining the Lehigh campus.

In bringing all research activities to a single location, Homer also decided that they should function as a single department which would be headed by a vice president of research. To that newly-created position, he appointed Charles H. H. Weikel, whose involvement with Bethlehem Steel's research program dated back to 1925 when he was transferred into what was then called the Patent Department. It later became the Development and Research Division, which is what most people think of as research: the scientific development of better products and better methods of production. From 1933 to 1957, Weikel served as manager of the Commercial Research and Industrial Development Division, a unit of the Sales Department which researched new markets, new customers and new uses for Bethlehem Steel's products. Commercial research is industry's crystal ball. It helps to confirm or refute gut assumptions. Sometimes it opens and defines a new avenue for business that was overlooked altogether.

Charlie Weikel became one of the top commercial research professionals in the steel industry. In 1944, while the war was still far from won, Weikel produced a far-reaching report titled "The Proposed Projects and Objectives for Bethlehem's Post War Expansion." Through elaborate graphs and statistics, this report plotted a strategic shift from Bethlehem's traditional concentration in heavy steel — construction, railroads and shipbuilding — toward lighter steel products, such as automotive sheets, household goods and consumer products. More than that, Weikel's report plotted geographic shifts in markets and converted all this data into a long and detailed list of capital investments to be made at each plant Bethlehem owned. To a large degree, this blueprint was followed. Its accurate predictions enabled Bethlehem to achieve significant growth after World War II.

In addition to that background, Weikel had served as assistant to the president from 1957 to 1960, the transition years when Homer was quietly planning Bethlehem's future. The fact that a career research executive served as Homer's right-hand man during that critical period surely had a lot to do with Homer's firm belief in the importance of having a strong research program.

The units consolidated under Weikel's supervision had been previously scattered under three other vice presidents. The Development and Research Division had been in the Steel Operating Department; the Commercial Research and Industrial Development Division had been part of the Sales Department; and the Raw Materials Research Division had been part of the Mining Department. Making those three departments give up their research turf, and, more importantly, getting them to cooperate across departmental lines with their former research units was a big accomplishment.

While Art Homer concentrated on making those sweeping organizational changes work, I concentrated on building the new $35 million research center.

When the planning of the research center began in 1957, Bethlehem Steel's research activities were scattered not only administratively but geographically. Different research activities were spread around at fourteen locations in four of our Eastern steelmaking plants. Because these were such diverse activities, and would remain so, our plans called for a cluster of eight major buildings attractively laid out across the 1000-acre mountaintop campus. These buildings would house such specialized research activities as process metallurgy, physical metallurgy, mechanical engineering, mechanics, ceramics, chemistry, chemical engineering and physics, including nuclear studies. In addition, the complex was designed to accommodate the advertising and marketing research activities as well as research administrative offices, cafeteria, auditorium, meeting rooms, photography labs, a boiler house, an electrical substation, and a 200-foot-wide circular cooling pond.

Our designers for the research complex were Vorhees, Walker, Smith and Smith of New York, who were probably the nation's top specialists in research center development. They had designed the famous Bell Telephone labs as well as research facilities for IBM, Ford, GE, Westinghouse, Du Pont and the French Government. When completed, Bethlehem had the finest, most advanced steel industry research facilities in the world.

The landmark feature of the research labs was the prominent green glass tower on top of the administration building. This became a prestigious meeting room for important conferences and other functions, and it gave the low-rise array of buildings a prominent and dramatic focal point that is visible from across the Lehigh Valley. Of course, the view of the Lehigh Valley from the Tower Room, as well as from almost anywhere on the research campus, is truly spectacular. A comment that I frequently made to visitors at the new research center was that Bethlehem Steel was "lifting metallurgical research out of the valley of traditional approaches to the solution of steelmaking problems and giving our research effort the range and perspective of distant horizons."

At the dedication ceremony on October 9, 1961, we named the new facility the Homer Research Laboratories in honor of the visionary executive who made them a reality. Ultimately, the complex grew to a total of eight major buildings, all of Indiana limestone in a contemporary design. The final price tag was around $70 million.

Homer Labs became well known throughout the Lehigh Valley and throughout the steel industry. Bethlehem Steel has a long record of technological innovation and leadership, and the Homer Labs became an outward symbol of that tradition. From its earliest days, Bethlehem Steel's innovations have had far-reaching effects, not only in the steel industry but in the affairs of the world. A generation before Charlie Schwab acquired the company, John Fritz established Bethlehem's unrivaled status as the world's foremost producer of armor plate and heavy forgings for ordnance and machinery. In 1898, two Bethlehem steelmakers, Taylor and White, discovered a process for making high-speed tool steel, an innovation which doubled the potential capacity of every machine shop in the world.

Another great example of industry leadership is of course the wide-flange beam, which for decades has been Bethlehem Steel's trademark. Invented by Henry Grey, the wide-flange beam became a commercial reality through Schwab's foresight. With further improvements developed by George Blakeley, the wide-flange beam soon became one of the 20th Century's most significant innovations, ushering in the skyscraper era. Blakeley later became president of Bethlehem's bridge-building subsidiary, also an innovator in its field.

Dr. Charles Herty was another Bethlehem innovator. As a pioneer in the application of physical chemistry to steelmaking reactions, his research led to many improvements to open hearth and other steelmaking practices. Later Bethlehem research made substantial improvements to the technique of injecting oxygen into the open hearth furnaces to speed the steelmaking process before the era of the basic oxygen furnace. And it was Bethlehem research that solved the problem of the shatter cracks which developed in steel rails when they were subjected to heavier car loadings. Bethlehem engineers developed a controlled cooling process that produced rails that would stand up under heavier axle loads and higher train speeds. It has been said that this product of Bethlehem research was yet another of our significant contributions to the winning of World War II, because railroad equipment could be used more intensively than would otherwise have been possible.

Bethlehem's mining and raw materials research people were the first to develop a successful method of transforming iron ore into concentrated pellets. We built the world's first pelletizing plant adjacent to our Grace Mine in Morgantown, Pa. Feeding pelletized iron into the blast furnaces greatly increased productivity, while lowering production costs. Other

Bethlehem innovations of years gone by include the method of pouring steel into the ingot mold in a vacuum in order to eliminate impurities from the finished steel product and Bethnamel, a patented vitreous enamel cold-rolled sheet which saves manufacturers of stoves and refrigerators the necessity of putting on two coats of enamel.

All of these breakthroughs and advancements were achieved by Bethlehem Steel before the establishment of Homer Research Labs. When the Labs were completed and the research department was reorganized, enlarged and installed in its new home, Bethlehem entered a new era of scientific leadership.

With our research organization in place at Homer, Bethlehem Steel was equipped to meet its ever-increasing challenges scientifically. This included new processes as well as new products.

Galvalume was one of the most successful discoveries developed at Homer. A sheet steel with a special zinc and aluminum coating that is far superior to plain galvanizing, this product is used world wide as a building material. It brought steel out of the architectural ugly-ducking stigma of the quonset hut and made it fashionable for the luxury construction market. Bethlehem Steel made millions licensing Galvalume technology to other steel companies.

By the time the last research building was completed in 1969, we had 900 people working up there. Later on, as the company was downsized, so was the research program. By the mid-Eighties, it was down to about 300 people. In 1986, Bethlehem Steel sold five of the buildings and about 700 acres to Lehigh University. Lehigh already was using some of the labs for its high-tech business incubator program, which is funded by the state's "Ben Franklin Partnership." Consequently, the state kicked in $10 million to help finance the purchase. The whole transaction was a good move — for the company, for Lehigh and for the community. Lehigh was able to greatly expand its high-technology and industrial research activities. Because of the long-standing friendships between Lehigh and Bethlehem Steel, both at the management level and among the scientific professionals, the Labs in many ways are still an important asset to the company. And, although many people don't realize it, Bethlehem Steel still has a significant research program — with about a $20 million annual budget — headquartered at the three remaining mountaintop buildings.

Even though the entire complex is no longer named for Arthur Homer (Lehigh calls its buildings the Mountaintop Campus. And they've named the big tower building Iacocca Hall after Lee Iacocca, class of '45, who is both an active alumnus and a generous contributor to Lehigh's high-tech programs) the facilities will always be the important center for scientific leadership and economic advancement that Art Homer envisioned.

1963, Art Homer's final year as Bethlehem's CEO, was a year of numerous other changes and milestones.

We saw a significant decrease in shipbuilding activity, which prompted the closing of our celebrated shipyards at Quincy, Staten Island and Brooklyn. It was a sad note on which to end Homer's career.

Steel imports in 1963 increased to 5.5 million tons, which was almost double what it had been just two years earlier. I think 1963 was the first year we made a pie chart to show that imports had become a noticeable part of the American steel market, growing from a mere sliver to a significant slice of the pie.

In 1963, Bethlehem Steel also began the construction of its first two basic oxygen furnaces. Scheduled for completion in 1964, the two 250-ton furnaces were being built at Lackawanna to replace 14 open hearth furnaces. Within a few years, BOFs would also be built at Sparrows Point, Bethlehem and Burns Harbor.

My promotion to the chairmanship of Bethlehem Steel occurred at a time of incredible changes — at Bethlehem Steel, in the steel industry, in the American way of life and in world economics.

1

1. Frances and me, in front of "Oak Knoll", our house in Saucon Valley. We put in a swimming pool and a practice golf hole.

Magazine Covers. As chairman of the 14th largest corporation in America, I went from newspaper headlines to magazine covers, including *Forbes*, *Iron Age*, and our own *Bethlehem Review*. That's Jake Jacobs with me on an inspection trip at Burns Harbor.

Chapter Nine

Chairman and CEO

In recent years, Bethlehem Steel and other large corporations have frequently appointed top executives to the somewhat mystifying and misunderstood position of "vice chairman."

At one point in the Seventies, Bethlehem had three vice chairmen. On the other hand, in the early Nineties, Bethlehem had no vice chairman and, more astonishing, nobody serving separately as president! Instead, the top management under Chairman (and President and CEO) Walt Williams consisted of an executive vice president and four senior vice presidents. There are many reasons for these creative arrangements. For one thing, running a big corporation isn't getting any easier. You have to build the best management team you can, without confining it to the typical corporate ladder configuration. For another thing, these deliberately unfamiliar pecking orders offer flexibility — if there is no heir apparent, you can have two or more candidates warming up in the bullpen. Responsibilities and turf can be shifted in subtle but strategic ways to strengthen the team's performance without resorting to a disruptive shake-up.

In this regard, I was a pioneer (or perhaps just a guinea pig) when I became the first vice chairman in Bethlehem Steel's history on August 1, 1963. On that same day, Stewart S. Cort succeeded me as president. These appointments, said Chairman Art Homer, were "a reflection of Bethlehem's policy of broadening the scope of the company's executive leadership."

This was characteristic of Homer's belief in the importance of continuity and smooth transitions. His own retirement was more or less slated to occur in less than a year, and he wanted to leave the company in good hands. There was never any hint that my appointment as vice chairman was a steppingstone to oblivion, a fate that sometimes occurs. On the contrary, this intermediate move enabled Homer to turn more responsibility over to me as I, in turn, handed over my presidential duties to Cort. Cort had come directly from the West Coast, and my presidency had been brief, so we both benefitted from this opportunity to adjust in stages to Homer's departure.

When Art Homer retired in April, 1964, he paid a high compliment to Cort and me and the new generation of executives who comprised our team when he said he was "confident that our management was never stronger or better prepared to face the industrial problems of the day." Homer, of course, deserved much of the credit himself for building that strong management team.

Two of the first things I did when I became chairman and chief executive officer of Bethlehem Steel were lower the executive pay and get outside directors on the board.

We were being severely criticized on both accounts, and I felt it was time to listen to our critics. Up until then, criticism came mainly from one or two oddballs who always showed up at the stockholders' meetings. But in the Sixties, the voices of dissent began to echo in respectable circles. We had to listen.

And it wasn't just voices of dissent. We were being sued by our stockholders because of the big bonuses we paid our executives. I asked our lawyers, "Why do we have all these lawsuits?"

And they answered, "Because you don't have any outside

directors, and you set your own pay."

Mr. Grace used to get a million dollars a year. He was by far the highest paid executive in America. At one point, I think Bethlehem had six of the top ten salaries in this country. So after I became chairman, we cut the bonuses way back, and at least we kept our names out of the papers.

As a result, I made far more as operating vice president than I ever did as chairman.

Next, I told the board, which was all company men, "We're going to have to get some outside directors."

The first one I went to ask to be our first outside director was Jack Rathbone, who was chairman of Standard Oil, now Exxon. I made a date and went to New York to see him, and he said, "I wondered why you were coming."

I said, "Well I came because you don't have any outside directors and neither do we; but we're going to have them."

He said, "I'll let you know in about a week." Well he did. Rathbone called me a week later and said, "I checked up with my people on this, and I want to do it." So he was the first one.

Then I asked Harold Helm from the Chemical Bank, Tom Gates from the Girard Trust in Philadelphia, Robert Hobbs, the chairman of First National Bank of Maryland, and Dr. Jess Davis, who was president of my alma mater, Stevens Institute. Altogether, we had five outside directors, and they came on board on January 1, 1965. A year or so later, we got Crowdus Baker, chairman of Sears Roebuck.

And we didn't have any more lawsuits over what we got paid.

Up until then, Bethlehem's policy also had been not to let any of our people serve as directors of other companies. I changed that, too. The first outside corporate board I joined was Morgan Guaranty Bank. I'm still on their advisory committee. I get paid $15,000 a year, and $575 extra when I go to a meeting, but I stopped going in 1990.

But as of January 1, 1993, they're raising my base pay to

$30,000 a year. Double what it was. At age 90! That's what I'll get paid!

My first year as chairman of Bethlehem Steel coincided with the hundredth year of the American steel industry. It had been in September, 1864, that the first commercial heat of steel was tapped by William Durfee at the Kelly Pneumatic Process Company of Wyandotte, Michigan. The process had been perfected experimentally a few years earlier by William Kelly at the Cambria Iron Works in Johnstown. Durfee's converter was based on the Kelly model, the original of which now rests in the Smithsonian Institution, a gift from Bethlehem Steel.

September 14, 1964, marked what we referred to as the beginning of "Steel's Century Two." On that date, more than 150 steel plants in the United States, Canada and Latin America poured commemorative 10-pound ingots from their first heat of the day. Appropriately, the very first of these to be poured — shortly after midnight — was at our Johnstown plant. From throughout the Western Hemisphere, these first ingots of steel's second century were then sent to the Battelle Memorial Institute in Columbus, Ohio, where they were remelted, with chrome and nickel added to the heat to produce stainless steel. The resulting ingot was recast into a sculpture of some kind.

Steel's Century Two observance also featured a nationwide news media tour of the steel industry. The first stop on the five-day tour was our new Homer Research Labs. After that, we showed them our new $16-million 12- and 14-inch bar mill at the Bethlehem plant. From there, the news people were flown around the country in a plane chartered by the American Iron and Steel Institute to see additional steel mills, iron ore mines and other facilities.

That new combination mill at the Bethlehem Plant was part of a $200-million improvement program during the mid-Sixties that also included two new basic oxygen furnaces and a lot of fancy stuff to control air pollution. We really modernized the plant and cleaned up the air coming out of it.

I think we successfully addressed our environmental challenges before they became mainstream issues on the national agenda. It was tremendously expensive, which further widened the gap between our production costs and those abroad. But by 1970, the year of the first Earth Day, Bethlehem Steel's emissions were clean, and nobody was attacking us for being polluters.

Changes in a steel plant move with the speed of a glacier, so the trick is to plan ahead and start making changes years before somebody tells you to. That's easier said than done.

There were many signs of change, both inside and outside of Bethlehem Steel, during my chairmanship: the social upheavals of the Sixties, the advent of steel's second century, and fundamental changes in the steel business and the economy. This combination of signals and trends meant that we had to think constantly about changes at Bethlehem Steel. There had been many long, happy years of growth and profits and a comfortable status quo. A job at Bethlehem Steel meant a life's career and a way of life. There was security and prosperity. But as changes took place around us, we had to change too.

I had to initiate quite a few disruptions of the status quo. I called upon every department head to find ways to reduce costs and eliminate non-essential expenses. They called it "Martinizing." The name came from a well-known local one-hour dry-cleaning service, but people thought it was sort of an appropriate term for some of the sudden and unpleasant things we had to do in almost every department of the company.

We were still making as much steel as ever, but because of such things as the decline in shipbuilding, the introduction of automation, ever-increasing personnel costs and an ever-shrinking profit margin, we had to, well, "Martinize" wherever we could. After two solid decades of post-war prosperity, many parts of the company were in need of streamlining. There were fiefdoms that had been built up — the stories have become almost legendary — and a lot of them did their jobs extremely well. It was just a matter of curtailing things we could no longer afford. Of course, a lot more has been done since then. Every CEO since me has had to tighten the belt a little further. Some of them, Don Trautlein and Walt Williams for example, have had to make cuts that far exceeded what I ever did. But nobody else ever had it named after them! Today, it doesn't need a name, because every well-run company has to have a permanent cost-reduction program.

I ran into Walt Williams one day not long after he became Bethlehem's CEO. It was the first week of the month, and I knew that a quarter had just ended, so I asked him, "How did you make out this quarter?"

He said, "We don't know. We won't know for a couple of weeks."

And I said, "Hell, we used to know on the fifth day!"

"Well," he said, "we laid off so many people it will take us about three weeks to find out how we did."

I don't know how many different layers he cut out of the management. One way to measure it, I guess, is to count the floors they still occupy at Martin Tower.

When we decided to build the tower, in about 1968, our people were in twenty-some different buildings around town. We even had an old hotel down near the coke ovens that we used as an office. We had, oh, maybe a couple dozen people in it. The tower has 21 floors, but now I think Bethlehem Steel only uses about half of them. They cut down to beat the band, but that was eight or ten years after I was gone.

When the tower was being planned, the company was still growing. In fact, the projections then were that the tower would accommodate additional growth only through the 1970's. After that, it was projected, even more space would be needed! Of course, that never happened.

I never had an office in Martin Tower! It was completed in 1972, after I retired. My last office was in the old general office building on Third Street in South Bethlehem, in the same sixth-floor executive suite that Schwab, Grace and Homer —

but mostly Grace — had occupied. Room 613. It's still there. Still looks the same, too. Same bathroom. Same venetian blinds. Same view of the plant.

One of the fixtures in the old executive suite was Millard Egbert. I still see him around town. Millard was one of the private secretaries brought in by Mr. Grace. He was there under Homer, me, Cort, Foy and Trautlein, and when they moved to the top floor of Martin Tower, he helped lay out the office. He could do everything: compose a letter, type it with twelve or fourteen carbons and sign your name.

Back in the old days, we never had anything but male secretaries in Room 613. Actually, Room 613 was several rooms — enough to hold the chairman, the president, the senior vice president and their various assistants and secretaries. And they were all men, until Ivor Sims (who became senior vice president under Cort and me) brought in Anne Maloney. She was the first of the new era, and Millard was the last of the old.

The other staff member in my office, along with Millard Egbert was Don Smith, who was my administrative assistant. He also came in during the Grace regime. Back in the days when Mr. Grace would spend a month or two down in Aiken, South Carolina, it was Don Smith who usually got summoned south so that Grace could have some of the trappings of Room 613 between rounds of golf.

When Homer became CEO, the office staff was no longer asked to transplant itself at the chairman's behest. The most exotic assignments, according to Millard, involved making arrangements for Homer's yacht-racing activities, such as the Halifax and Bermuda races.

In spite of "Martinizing," Bethlehem Steel in the Sixties was still a huge organization with far-flung interests and installations that took us to the four corners of the globe.

One of my most fascinating projects as Chairman was our LAMCO-Bethlehem iron-ore venture in the Nimba Mountains of Liberia.

The search for iron ore has always taken steelmakers to remote and exotic locations. Young Eugene Grace was given his first career-proving assignment in 1906 when Charlie Schwab sent him down to Cuba to reorganize Bethlehem's iron ore mines in the jungle highlands. (Grace's only fear was that he might be stationed down there permanently. So he made Schwab promise to bring him back if he got the job done.)

The primary concern in exploring and developing new sources of ore is to guarantee a long-term future supply of the highest-grade ore possible. Castro's Communist takeover in Cuba effectively cut off that source of ore, and this prompted us to pursue the Liberian venture.

The Nimba Range is located in the dense jungle highlands near the west coast of Africa, about 160 miles from the Atlantic Ocean. Nimba is reached by travelling up the St. John River, which roughly bisects Liberia, from the port of Buchanan on the coast (south of Monrovia) to the town of Gba. Nimba is farther inland. Exploration begun in the middle Fifties had ascertained that the Nimba Range contained some 235 million tons of proven high-grade 66-per-cent ore, with the probability of far more in unproven reserves. The project was launched in earnest in about 1960.

Bethlehem Steel entered into a quarter-billion-dollar joint venture with the Liberian government and other international interests. Bethlehem's interest was 25 per cent, and the remaining 75 per cent was owned by LAMCO, the Liberian American-Swedish Minerals Company. Half of LAMCO, in turn, was owned by the Liberian government, and the other half was owned by a foreign consortium organized by the Wallenberg banking family of Sweden.

The vast project was typical of those that draw upon virtually all of Bethlehem Steel's many areas of expertise. In addition to establishing the mining operation itself, a 165-mile railroad had to be built from Buchanan to Nimba, including two bridges over the St. John River. Bethlehem's Steelton plant furnished all the rails and track accessories, such as switch stands, frogs, guard rails, joint bars and tie plates. Rolling stock included diesel locomotives and 90-ton flat-bottomed ore cars. A deep-water port had to be developed at Buchanan to handle 45,000-ton ore carriers, and modern communities had to be developed both at Nimba and Buchanan to house and accommodate the work force.

Among the efforts to create a civilized community at Nimba, the partnership established a small country club, which showed movies twice a week and boasted Nimba's first tennis courts and football (soccer) field, not to mention the inevitable golf course.

The project had the enthusiastic backing of Liberia's president, William V. S. Tubman, a man whom I grew to know and admire. Tubman was a strong advocate of the industrial development of his nation, and he foresaw the wide array of jobs and business opportunities that would spin off of the LAMCO venture. By 1963, LAMCO was shipping six million tons of ore per year out of Nimba, an amount that eventually doubled.

One of the most memorable honors I've received in my life was an award Mr. Tubman conferred upon me in December, 1966, for my role and that of Bethlehem Steel in leading and guiding the economic development of Liberia. One of Liberia's oldest and highest distinctions, it is the Grand Band of the Order of the Star of Africa, which is roughly the equivalent of Knighthood in a European country.

As a matter of fact, I was knighted by the Swedish Government because of what we did in Liberia. They made me a Knight Commander of the Royal Order of the North Star.

So I'm a knight on two continents!

Never a dull moment!

In 1965, Frances and I were in a serious automobile accident on the German Autobahn. We were in Germany with Floyd Eckhardt, our chief engineer, and Jorge Schneider to inspect a coke oven plant. At the time, Bethlehem was building our new steel plant at Burns Harbor, but we hadn't built the coke ovens yet. A method had been perfected in Germany for building coke ovens twenty feet high. All coke ovens in this country were only fourteen feet high. It's one thing to build a bigger coke oven, and another thing to build one that works. So we were anxious to see how they worked.

Sure enough, it appeared that the larger coke ovens could successfully operate, even with a 50% bigger load of fuel. Our trip therefore had been very worthwhile, and we were going to drive down to Italy with Floyd and Betty Eckhardt in a big rented Cadillac. The day after we toured the coke ovens, at nine in the morning, we had this terrible collision.

We were going about 90 miles an hour. Up ahead of us, a big truck carrying a load of new cars had hit an overpass. There wasn't enough clearance for the cars up top, so the rig was all smashed up. I was sitting up front, and I saw it quicker than our driver. In fact I was just turning to get his attention when I went through the windshield. I had a felt hat on, and it must have kept me from being seriously cut. My main injury was a shoulder separation.

Frances and Betty Eckhardt were hurt far more seriously than me. Betty Eckhardt's front teeth were knocked out, so she was all bloody. And she was knocked unconscious for more than half an hour. I thought she was dead. Floyd had landed on top of both wives. He cracked his jaw when it hit the cigarette ashtray, but he would never admit he had been hurt.

At first, I was walking all around the accident scene trying to be helpful. Then I realized I was injured and risked getting hurt even worse. I thought to myself, "You damn fool! Stop walking around." We just had to wait for help to come.

We take for granted the speedy response of the police and other emergency services on American highways. On the Autobahn — and this was more than 25 years ago — it took forever to get help. Apparently, there was another accident a couple of miles ahead of us and several people were killed. So they took care of them ahead of us. We didn't get taken to a hospital and get looked at until three in the afternoon. It was a small-town hospital, and I was concerned that they weren't equipped to give Frances the care she needed, so I got us transferred to the U.S. Army hospital in Weisbaden.

I also got to a phone and called my office. I couldn't get my secretary. I talked to Lew Foy, then an assistant vice president. I still didn't know how seriously we'd all been hurt, but I told him I had a shoulder separation at least. The Army hospital of course had assigned doctors — they just happened to be there as their assigned duty, so it wasn't like a private hospital where the good doctors get promoted. I didn't want an assigned doctor. I told Lew Foy I wanted my own doctor, Dr. Dziob. He had been the company doctor up at the Lackawanna plant, and he was good.

Meanwhile, the Army doctor told me he was amazed I wasn't more seriously hurt. "You're lucky," he said. "You don't even have a concussion." (Maybe it was my hat.)

Dr. Dziob arrived in Germany the next morning. He put casts on my shoulder and my ankle, which also was broken. It turned out Frances had a cracked pelvis — cracked in four places. I hired a private nurse for her, and Dr. Dziob told us she'd have to walk every day so that she would heal properly. The very first day, he said, "Mrs. Martin, I don't care if you cry, but I want you to stand up and walk." It was agonizing, but Frances did it.

In a day or two, Dziob got me discharged from the hospital. We'd been on our way to see another steel plant in Italy at the time of the accident, so I flew down there about a week after the accident.

When she was able to be moved, Frances came home to America on a stretcher. She was stretched across three seats on the plane, and when we got home to Bethlehem, she was in St. Luke's Hospital for about a month. After that accident, sitting was always uncomfortable for her. She used to remark that the worst part of it was she couldn't even sit to play bridge.

Another international adventure on behalf of our quest for improved steelmaking took us to Spain. One of the ingredients needed to make steel in a basic oxygen furnace is the mineral Fluorspar. It helps to separate the slag from the steel. It turned out that there were abundant Fluorspar reserves in northwest Spain. Bethlehem was going to develop the Fluorspar mine in cooperation with the Spanish government, so Frances and I went to Spain to meet with the various people involved, inspect the operation and pay a courtesy call on Generalissimo Franco.

We were accompanied throughout the trip by Ed and Lala Leach. Ed was in Bethlehem's mining division, and his career had included a number of overseas assignments. It was while stationed at our iron ore mines in Chile that he met Lala, who had moved there with her British parents. Not only is Ed a top mining engineer, but the two of them are very well versed in the subtle areas of international business diplomacy and protocol.

Our host for most of the Spanish visit was the Duke of Arion, a charming and sophisticated man whose aristocratic family dated back many centuries. He had extensive personal contacts and the ability to open doors and get things done. In old-world cultures, a person like that is critical to the success of a new venture, so the Duke of Arion was some sort of consultant or paid advisor to us. Of course, he and the Duchess — she was a Hapsburg from Austria — very graciously hosted us at their country estate and saw that we were lavishly entertained wherever we went.

Among his various talents, the Duke was a champion sailor. He could have competed in world-class events like the America's Cup. However, he said he wouldn't compete against the United States because he felt we spent too much money.

We waited until the final day of our trip to have an audience with Franco, who'd been away taking a vacation. To go to a meeting at Franco's palace, you had to dress in a formal morning suit, with a swallow-tail coat and striped trousers. I had to go to some Spanish tailor and rent an outfit. He had trouble finding clothes in my size, because I'm not built like most Spanish people are. Well, he finally came up with a coat that fit me, but he didn't have any pants even close to my size. So he took the biggest pair of pants he had and sewed a big wedge of cloth into the seat. The patch didn't even match the pants material; in fact, it was sort of a red striped patch. But that wasn't going to matter since it would be hidden by my tails. The Leaches have movies of our entire trip, and the best part of their movies is when they got a shot of my patched rear end when I was getting on the plane.

Fortunately, Franco never saw the patch or the movies. We did have a memorable meeting. I was told by the Duke of Arion that when Franco had foreign visitors, he often promptly went to sleep in their presence. On this occasion, he'd just gotten back from a salmon-fishing holiday, and I got to talking with him at great length about salmon fishing. Franco and I ended up having one of the longest palace audiences the Duke had ever seen!

We also had lunch with Prince Juan Carlos while we were there. He'd been designated by Franco to become King of Spain, which happened a couple of years later when Franco died. His wife, the Princess, was very glamorous. At the time, she had the Jackie Kennedy look.

While we were fascinated by the old-world social customs of Spain, we were shocked at their lack of progress in environmental protection. Of course, pollution control was very much on my mind throughout the Sixties. But in Spain, they were not yet serious about it. The Fluorspar mine was near a beautiful beach on the northern coast of Spain. They were really wrecking that beach developing the mine, but neither the local nor central Spanish government was concerned. Industrial development was more important. They also had to tear down a historic church near the mine. There was no battle or protest, like you'd surely have in this country.

Not all my assignments as Bethlehem Steel's senior representative took me to exotic foreign lands. There was plenty to do in this country. We had major company facilities in numerous states; we had financial business in New York; and we had political business in Washington.

When I got to be head of the American Iron and Steel Institute — which is all the steel manufacturers and research people, and usually the chairman of it is somebody who's the head of a major steel company — we had tough competition amongst the steel companies and internationally, too. And our competition from non-steel products was growing, too. It was a triple-threat. This was in the mid Sixties, during the Johnson Administration.

Part of my responsibility as head of the AISI was to convince the people in Washington that our biggest problem was imports. England, for example, would ship steel into this country, and they were losing $200 a ton on it, and their government would make up the difference. A private company can't compete with a foreign government!

Roger Blough, the head of U.S. Steel, and I, by talking to people in Washington like Wilbur Mills, got the first voluntary import quotas established. We used to go down to Washington and we'd have a date with Wilbur Mills for 8 o'clock in the morning.

My idea of being on time is to be at least half an hour early, so I'd arrive in Mills' office a little after seven. We'd have a cup of coffee together before the meeting, and that informality helped me to convince him to go to bat for us. He got the real Ed Martin, and I got the real Wilbur Mills, before anyone else showed up. And it worked. Mills unfortunately was having difficulty at the time with his personal life; he got in trouble, you know, when they caught him with some girl swimming in the Tidal Basin. Too bad. He was a hard working guy. He evidently was an alcoholic, and he resigned after that episode, but he helped us to get the voluntary quotas through.

I met with Lyndon Johnson once when he came to the shipyard at Sparrows Point for something. It may have been the time we had Ladybird christen the Navy ship "Versailles." I remember being annoyed because all Johnson wanted to do was talk to reporters about everything but the business at hand.

When I met Richard Nixon a few years later, it was at the Presidents' Club in New York. It was for presidents of corporations, but since he was the president, he wanted to be a member. The club has kind of a fun tradition. The first time you come to a meeting, you have to give a speech on how you got to be president. But, to your astonishment, everybody in the audience laughs on purpose at everything you say! It sort of breaks the ice. I remember we did that to Nixon when he joined.

I remember going to Japan when I was head of the American Iron and Steel Institute. They'd rebuilt their war-decimated steel industry and they were already shipping a lot of steel to America. I made a lot of speeches to them, and they smiled and printed all the speeches, but they kept right on doing what they were doing. You couldn't trust them when they smiled. You still can't.

In spite of the growing import problem, Bethlehem made out all right during the Sixties. We competed with all the different companies and did as well as you could expect. Our annual output stayed high, and we were still profitable.

One of our biggest strategies during that period was to try to capture as much of the auto industry's sheet steel business as we could. I talked to all the top people at General Motors, Ford and Chrysler. I had to go around and be a salesman to people like Henry Ford — the grandson — who I thought did a good job running his company.

I was a member of the Augusta National Golf Club so that we could entertain customers down there. We'd rent a house and hire a cook and invite different customers to come down and stay with us for the national tournament. I remember playing golf one time with the chief buyer from GM and the chief buyer from Chrysler in the same game! I forget who our fourth was.

We also kept up a lot of our structural business in those

days, though our labor costs were pricing us out of the market. We had a lot of big structural jobs in the sixties. We built the George Washington Bridge across the Hudson at 125th Street. It cost as much to put the second tier on 12 or 14 years later as it did to build the original bridge. We built the whole bridge.

On the Verazzano Narrows Bridge, we built one tower. Our steel is in the other tower, but the prime construction contractor wanted to build it, so we sold them the steel and they built the other tower. U.S. Steel built the approaches to the bridge.

Bethlehem Steel had been in the bridge fabricating business since the Thirties, when Grace bought the McClintic-Marshall company in Pittsburgh. McClintic and Marshall were classmates at Lehigh about ten years ahead of Grace. First he bought their company, then he convinced them to donate a lot of their money to Lehigh!

Shortly after that transaction, the McClintic-Marshall subsidiary was building the Golden Gate Bridge in San Francisco. On an inspection visit, Mr. Grace asked one of the steeplejacks who was building the bridge, and the guy said, "McClintic-Marshall." Grace did away with the name right after that, because he wanted people to know Bethlehem had built the bridge.

Bethlehem eventually went out of the fabricating business because, at the time, I think we were paying something like three dollars an hour and our competition was paying only a dollar an hour. We were paying steelworkers' wages, and they weren't. So we told the union, "We're going to get out of this business unless you give us decent labor costs so we can compete."

The union didn't believe us, so we went out of the fabricating business. We closed our Pottstown plant where we did most of the bridge fabricating. The whole town nearly went to bed after that.

Our shipbuilding division was slowed down for two reasons: airplanes were taking a lot of the shipping companies' business, and the extended Cold War meant that nobody's navy vessels were being sunk. The demands for replacement and repair were greatly reduced.

But there was one part of the maritime industry that was thriving during the Sixties. The oil shipping business was growing every year, due to increasing demand, increasing prices and increasing inconveniences, like the closing of the Suez Canal.

One of our biggest shipbuilding customers was Aristotle Onassis. He hadn't married Jackie Kennedy yet, but he was at the height of his power as a global oil shipper. We built several super-tankers for him.

I remember getting a letter inviting Mrs. Martin and me to have dinner on board Onassis' yacht on the Hudson River. A barge took us out to this big yacht, the "Christina." It had a big tiled swimming pool, which had a roof that could be closed over it, and that's where we ate dinner. There were about two dozen of us. I remember David Rockefeller was there, and other prominent business tycoons. Frances sat next to Onassis, and I sat next to his girlfriend, the opera star Maria Callas. The menu was printed in French, and she ordered for me.

After dinner, Onassis took me aside and said, "Mr. Martin, I want your advice. I want to provide a steel plant for Greece, near Athens. Greece has no steel industry. What do you think?"

I told him not to do it. There was already too much excess steelmaking capacity in the world. I told him, "If you want to help Greece out, just give them money."

After that, until he died, every Christmas I got a case of Moët Chandon Champagne. All that was inserted was his business card.

My world travels on behalf of Bethlehem Steel were fascinating, but I always was just as interested in all the different things that were going on right in and around the Bethlehem plant.

Bethlehem Steel in those days had a widespread policy of doing all kinds of far-flung tasks in-house. We had all these little businesses and specialty services, so we wouldn't have to go out and hire a company every time we needed something. We were almost totally self-sufficient. Unfortunately, I had to get rid of some of these operations when I was Martinizing, and more were eliminated in later years. But in the Sixties, those operations were all part of Bethlehem Steel's vast size, strength and versatility. One of the big in-house operations that I found especially fascinating was the pension fund investment program.

While I was chairman, the manager of Bethlehem Steel's pension fund was Dave Dunbar, a Harvard graduate who had joined the Loop course in 1940. According to Dave, the pension fund grew from $30 million when it was created in 1928 to around $2.5 billion when he retired in 1983. This big portfolio of investments was managed by Dunbar and a staff of six. They were as sophisticated as a small brokerage house, initiating timely trades and transactions and maintaining mountains of analytical data — annual reports, Wall Street publications and so forth. The average size of each investment transaction was around $5 million. They did everything but actually execute the stock transactions, which you can't do without a brokerage license. For that, they dealt with about sixty licensed brokerage houses. Every so often, at directors meetings, for example, someone would ask why we didn't just turn the whole job over to Morgan Guaranty Trust or some other outside investment manager. Whenever the question was raised, Dunbar's people were always able to show that they were doing the job as well, if not better than an outside firm — and at half the cost.

The work of Dunbar's staff was overseen by an investment committee composed of about half a dozen of the company's top officers and financial men. They used to meet the first Monday of every month. When I was Chairman, I liked to drop in on those investment committee meetings whenever I got the chance. I went for two reasons: first, I thought the sound management of two-and-a-half billion Bethlehem Steel dollars warranted some attention from me; second, it was an activity that fascinated me. Dunbar claims that my appearances at those investment meetings always gave an added charge to their deliberations. I enjoyed listening as our financial experts made strategic investment decisions, and they enjoyed (or at least pretended to enjoy) my occasional tidbits of information about the various Fortune 500 business leaders I encountered in the course of my job.

I enjoyed working with my executives, especially when the circumstances were informal. Unfortunately, being CEO of a company as big as Bethlehem Steel means you frequently are expected to operate in very formal settings and circumstances. The bane of my existence was making speeches. I had to make speeches all the time. When I retired, they gave me a bound book of all my speeches, and it's about four inches thick.

Fortunately, I had outstanding help. The fellow I relied on most was Larry Fenninger, our assistant vice president for public affairs. His background was in labor relations, and there was some occasion when he was getting together a lot of information I needed for a meeting in Pittsburgh or somewhere. I was telling him what I wanted, and he was sitting there taking notes, so I said, "Larry, why the hell don't you just write it for me."

After that, Larry Fenninger wrote all my speeches. Up to then, he'd never written a single speech.

At one point, the company hired a speech instructor from New York City. She must have had a career on the stage, because she was like something out of a Wagnerian opera. Her name was Maxeda Von Hess. And she weighed around 300 pounds. She coached me on my delivery. Sometimes she would just change one word, and I could read the whole speech without stumbling.

Another of Madame Von Hess' secret techniques was to have me take deep, rhythmic breaths for four or five minutes before giving a speech. Needless to say, there weren't too many occasions where it was feasible to do that before making a speech! In any event, the technique ceased to be a secret after the Wall Street Journal did one of their tongue-in-cheek stories on Ed Martin's public-speaking coach.

Among my speechmaking responsibilities, I had to chair the annual stockholders' meetings. We always held them in Wilmington, because we were a Delaware corporation. When I became chairman, I didn't like the way we were running the meetings. We always had a couple of trouble-makers, and the meetings would end up on a negative note.

I decided to see how some of the other corporations ran their meetings. I went to an Exxon meeting up in New York; they must have had a couple of thousand people there, and it ran like clockwork. Then I went to a DuPont meeting. There were about 800 people at that meeting, and afterwards, I asked the chairman, Crawford Greenawalt, why he had so many people attending.

He answered, "Eighty percent of them are members of the family. You can't keep them away!"

The advice Greenawalt gave me was to have a long program. Have both the president and the chairman make long speeches. The people are only going to stay a certain amount of time anyway, so you'll have less time to have to answer questions. We did that, and our meetings were much better after that.

These were some of the responsibilities of the chairman. They had nothing to do with making steel, just with running the company.

Between the activities at the research labs, the new technology we were putting in, and things like pollution control, there were a lot of exciting challenges in Bethlehem while I was chairman. But the most exciting challenge of all was a project we were hatching out in the middle of northern Indiana, on the shore of Lake Michigan.

1

2

1. On our trip to Spain in search of Fluorspar.
2. Floyd Eckhardt and the Duke of Arion were my advisors in Spain. We are examining a model of our future mining site.
3. We flew home to America on the Bethlehem Steel jet.

3

1

1. Global affairs. Receiving the Grand Band of the Order of the Star of Africa from William Tubman, the president of Liberia, in 1966. We formed a joint venture to develop the Nimba iron ore range, which significantly improved Liberia's economy.
2. During the Johnson Administration, I was frequently in Washington on business. In 1964, LBJ and I signed the "Plan for Progress", establishing Bethlehem Steel's commitment to affirmative action.

2

1

1. Ladybird and Frances seem more interested in my speech than LBJ. The occasion was a luncheon following the christening of a new ship by Ladybird at the Sparrows Point Shipyard.

2 and 3. Both LBJ and I were the subject of editorial cartoons regarding Bethlehem Steel's significance during the Vietnam War.

2

3

"Plastic Planes? Plastic Tanks? Aw, Come On Lyndon!"

1

1. With Malcom Forbes aboard the "Highlander."
2. Frances and I and another couple arriving at the party celebrating the 50th anniversary of *Forbes* magazine in 1967, held at Forbes' estate in New Jersey.

2

1

1. The Gary Medal is the highest honor in the steel business. In 1968, the year I was president of the American Iron and Steel Institute, I presented the medal to Roger Blough, chairman of U.S. Steel, with Frances and Mrs. Blough in attendance.
2. Vice President Humphrey attended the same AISI dinner in 1968. Flanking him are Roger Blough, AISI president John Roche, me, and Thomas Patton, the chairman of Republic Steel.

2

1

1. Our rented Cadillac, after the crash on the German Autobahn. I went through the windshield and Frances suffered a broken pelvis. We were in Germany inspecting their new 20-foot-high coke ovens, a technology we hoped to employ at Burns Harbor.
2. We brought Frances home to the U.S. on a stretcher.

2

Chapter Ten

Building Burns Harbor

If I had to select one thing for my epitaph, it would have to be Burns Harbor.

There has only been one occasion in this country in the last forty years when anybody ever attempted to build a fully-integrated modern steelmaking plant — entirely from scratch — and Burns Harbor is it. Built during the decade of the 1960s, Burns Harbor still bears the distinction of being America's newest fully integrated steel plant.

The American steel industry has been criticized in recent years for having old-fashioned and out-dated steelmaking facilities. To a large extent, that criticism oversimplifies the issue. It's true that most American steel plants date back to the Nineteenth Century. Some, like our Sparrows Point, Steelton and Lackawanna plants, were built in the earliest days of modern steelmaking. Others, like the plants in Bethlehem and Johnstown, date back even earlier, to the dawn of the industrial age, when they were known respectively as the Saucon Iron Company and the Cambria Iron Works. It was at Cambria, in Johnstown, that modern steelmaking was invented in 1857, when William Kelly perfected his Kelly Converter.

Operating at the same site for five or six generations is a great distinction if you're a bank or a department store or a family farm. But it's a constant handicap in the steelmaking business. The heaviest of heavy industries, steel plants can't be moved easily. To far greater degrees than in other industries, you're restricted by cost, geography, zoning and many other factors. So you stay put and make do.

That's not to say that Bethlehem Steel and other leading steelmakers didn't modernize. Most of our plants have been rebuilt several times over, at a cost of hundreds of millions of dollars. The equipment may be state-of-the-art, but you're still squirming, gerry-mandering and leap-frogging around on a cramped site that was selected in an age that predated computers, telecommunications, global competition, the environmental movement, labor relations, the automobile and even electricity!

By the middle of the Twentieth Century, when most foreign steel industries were being rebuilt amid the rubble of World War II, American steelmakers (who in some cases were helping to design the new foreign plants) began exploring the possibilities of building anew. By and large, however, the challenge of modernizing and upgrading their existing facilities consumed the lion's share of their available investment capital, design capabilities and risk-taking initiative. For example, between 1945 and 1960, Bethlehem spent about $1.5 billion on improvements and expansion of capacity at our existing plants. We were producing steel at an unprecedented rate in order to supply pent-up post-war demand both here and abroad.

The only two American companies to build brand new steel plants after World War II were U.S. Steel and Bethlehem. U.S. Steel's new plant was built at Morrisville, Pennsylvania, just up the Delaware River from Philadelphia, and was named the Fairless Works after their chairman, Benjamin Fairless. This location gave U.S. Steel a Mid-Atlantic tidewater plant with which to compete with Bethlehem's Sparrows Point plant.

It was the same kind of geographic strategy that led Bethlehem to build at Burns Harbor, Indiana. A new plant there would enable us to compete with U.S. Steel and the other companies who dominated the growing Midwest market. It was in many ways a reciprocal move. Up until the Fairless Plant was built, there had been sort of an understanding that U.S. Steel and Bethlehem wouldn't invade each other's territory. But, when U.S. Steel built at Morrisville, Bethlehem deemed it imperative to establish a presence in the Midwest.

Bethlehem Steel had long sought ways to increase its access to the growing midwestern market. Other steel companies enjoyed a major presence amid the automotive, machinery and other heavy manufacturing industries of the region: U.S. Steel had plants at Gary, Indiana, and South Chicago; Inland was also on Lake Michigan; and Republic Steel was in Cleveland. Bethlehem, of course, had acquired the Lackawanna plant in 1922, which provided direct access to the Midwest by way of the Great Lakes. Nevertheless, Bethlehem's management wanted something bigger, better and closer to Chicago, the epicenter of the Midwest.

In the Twenties, Mr. Grace foresaw the growth in the Chicago market. The method he chose for entering it was the successful strategy he had been employing relentlessly since before World War I: mergers and acquisitions. Mr. Grace therefore proposed a merger with Youngstown Sheet & Tube, which already had steelmaking facilities in Chicago. The proposal was fought by Cyrus Eaton of Republic Steel, who was also eager to dominate the Chicago market. The conflict remained tied up in court until 1934. By the time all the obstacles were cleared away, the Great Depression was in full swing and the entire American steel industry was operating at about 40 percent of capacity. Profits had been wiped out, stock values had plummeted, and the incentives for acquiring Youngstown had virtually evaporated.

Two decades would pass before Mr. Grace resumed his quest for Youngstown. He sought another merger in 1954, but this time the U.S. Justice Department blocked what it deemed to be an antitrust violation. Bethlehem battled this position relentlessly over the next several years, arguing that the merger would increase, not decrease, competition in the steel industry. I think Mr. Grace was prepared to take it all the way to the Supreme Court, but he suffered his first stroke in 1957; in the very same week that Bethlehem filed a brief in U.S. District Court seeking a trial and decision, the company announced Mr. Grace's election as honorary chairman and the transfer to President Art Homer of the duties of chief executive.

After forty years of building up Bethlehem Steel through astute mergers and acquisitions, Mr. Grace saw that the prevailing winds of capitalism had changed for good. "Bethlehem couldn't merge with a blacksmith shop," was his oft-quoted response.

It's entirely possible that the frustrating Youngstown affair — the only unsuccessful venture of his career — hastened the decline of Mr. Grace's health. This all amounted to a rather historic and distinct turning-point for Bethlehem Steel, the end of an era and the beginning of another.

As the final curtain was coming down on the long-running Youngstown drama, an unknown site by the unlikely name of Burns Ditch was already waiting in the wings as the unannounced leading contender for the lead role in Bethlehem's Midwest quest.

The Chamber of Commerce of Valparaiso, Indiana, had first publicized the Burns Ditch property as a possible steelmaking site in 1930. Located in Porter County on the southern shore of Lake Michigan, the location was also being promoted by the Indiana state government as a future deep-water public harbor. The harbor was intended to serve the envisioned steel plant as well as other Great Lakes shippers doing business in the region.

Bethlehem started quietly acquiring acreage at Burns Ditch in 1956. The man responsible for assembling the real estate was George Hurd, who worked almost exclusively on this undertaking for the next six years. George recalls that the search for a new steel mill site began as early as 1950. Two locations on Lake Michigan were found to be feasible. One was Burns Ditch — that's what they called it before the harbor went in — not many miles east of Chicago. The other was west of Chicago, on the west shore of the lake. According to Hurd, the Burns Harbor site was picked by Charlie Weikel, who had marketing statistics proving it was the better site.

The purchase entailed some 150 individual parcel transactions. George Hurd had examined every square inch of Burns Ditch so many times that he knew the land far better than the owners did! One 900-acre parcel had recently been acquired by the Murchison family of Texas as part of a big package deal, and they really had no idea what they wanted to do with it. George had to wait for months without pestering them for a decision; he wanted to be certain that they had reached their own conclusion that they had no use for the property.

Building a steel plant in the second half of the Twentieth Century proved to have many modern challenges that required new responses. One was the growing environmental movement.

Some people say that Bethlehem Steel actually started the environmental movement, at least as it relates to industry, on the day Mr. Grace made his famous "bird sanctuary" remark to a reporter. This had to have been in the mid-fifties. When asked to comment on the rumors that Bethlehem Steel was buying up an expanse of sand dunes and marshes on the edge of Lake Michigan, Mr. Grace snapped, "We aren't acquiring it for a bird sanctuary!"

With Mr. Grace's typically blunt and forthright remark out in the open, at least Bethlehem Steel could proceed to interact with governmental bodies, citizen groups and the news media from a position that was straightforward and trustworthy. In a way, it was helpful that the famous quotation could be attributed to Mr. Grace, because those of us who built Burns Harbor could say, "There's a new generation in charge now." As it turned out, we went to great lengths to address environmental concerns, and Burns Harbor achieved a positive reputation in this regard, even as public opinion in the Sixties continued to mount in opposition to heavy industry throughout the country.

It had to be pointed out that only four of the 15 miles of lakefront duneland in Porter County were zoned for heavy industry and Bethlehem's 4,000 acre tract fell within this zoned area. Most of our tract contained no duneland at all, being composed mostly of submarginal farmland and swamp. Furthermore, our property had been in industrial ownership for almost 30 years, having previously been owned by a quarrying concern.

Construction at Burns Harbor began in 1962. Mr. Grace had been dead more than two years, and the old guard who'd served him was pretty much dead or retired as well. The team that normally would have implemented another one of Grace's quarter-billion-dollar initiatives was gone. By '62, there was an entirely new team in place. Art Homer was chairman of the board, and I was president. Homer appreciated the fact that I'd spent all of the last four decades in steel operations, so he gave me responsibility for the Burns Harbor project.

What an assignment! Building a steel plant from scratch was a once-in-a-lifetime opportunity, and fate had placed that opportunity on my desk. Like any steel operations manager who'd spent his career in plants that were older than he was, I was full of notions about the design and construction of the ideal steel mill. Of course, we had an outstanding team of engineers, planners, steelmakers and builders assigned to the project.

Among the team we assembled for the Burns Harbor project were some of my old staff members from Lackawanna. Floyd Eckhardt (who was in the car accident in Germany with us) had been my chief engineer and then assistant general manager at Lackawanna; he later was my assistant vice president when I was v.p. of operations, and ultimately he became Bethlehem's first vice president of engineering. Floyd played a major role in the Burns Harbor project. Another was Jim Dague, who had come to Lackawanna out of the Loop Course in 1939. During my tenure there, he was head of Lackawanna's mechanical department. In 1961, we moved him to Steelton, where he served as general manager before we sent him out to be the top man at Burns Harbor. In his capacity as Burns Harbor's general manager, he not only supervised construction, but also had responsibility for the operating departments as they came on line during the early Sixties. Dague's assistant general manager was Bill Ritterhoff, who later served in two of my old jobs as vice president of steel operations and corporate vice chairman.

For our chief engineer on the Burns Harbor project we chose Dick Hurd, who had served as chief engineer at Lackawanna during the latter part of my tenure there. In a typical steel plant organization, the chief engineer runs the non-manufacturing side of the business. He's in charge of all construction, renovation and equipment maintenance, and he reports directly to the general manager's office. Of course, when the primary activity is construction of the plant, his job is that much more important.

One of the engineers who contributed a great deal to the success of the Burns Harbor project was a young fellow named Walt Williams, another looper whom I'd hired at Lackawanna a decade earlier. I'm glad that fate also placed this future CEO of Bethlehem Steel in the midst of what was widely being hailed as the industrial project of the century. I know that today Walt is as proud of Burns Harbor as I am.

Burns Harbor was built in stages. In its first phase of development, it was not a fully-integrated, that is to say, "steelmaking" steel plant. The first phase contained only finishing facilities, including a 160-inch sheared plate mill, which went on line in December, 1964. Within a year, when we held our grand opening, the cold-rolled sheet and tin mills were also in operation, and an 80-inch hot-rolled sheet mill was nearing completion. These finishing mills were supplied with semi-finished steel produced at Lackawanna and Sparrows Point.

Part of the reason for our delay in establishing steelmaking facilities at Burns Harbor had to do with a profound technological change that was taking place at that time. The basic oxygen furnace was beginning to replace the open-hearth furnace as the standard steelmaking mechanism. The most significant difference between the two technologies was in the time it took to make a "heat" of steel. Whereas an open hearth takes eight hours to produce a 400-ton heat, a basic oxygen furnace (BOF) can make 200 tons of steel in 45 minutes. You save about $5 a ton.

Invented by the Austrians in 1952, the first BOF in North America was built in Canada a few years later. As with any new technology, the BOF went through a rapid series of refinements, the most important of which was making it large enough to make steel in high volume.

Of course, there is also a very basic, common-sense reason for building steelmaking facilities last. If you built steelmaking facilities before you had any finishing mills, you'd be stuck with semi-finished steel. By building the finishing end first, you can start making money, selling finished products, establishing a customer base and building a work force. To wait until an entire new billion-dollar integrated steel plant is completed would keep far too much capital tied up in unproductive facilities.

The grand opening of Burns Harbor, a week-long extravaganza in September, 1965, was a feat of showmanship that would have made Charles Schwab and Eugene Grace very proud. We spent a million dollars on that grand opening.

We began on September 16 with a "press preview" for about 200 members of the nation's news media who were flown in from New York, Washington and just about every place that Bethlehem maintained facilities or offices of some kind. Unlike unveiling a new automobile, this was a long and complex story that we were trying to convey. We had to be prepared for the difficult questions that would come from reporters who routinely covered the steel industry. At the other end of the spectrum, we had to provide a lively, writable, readable story to the general media representatives, which included *Time*, *Newsweek*, television and the major daily newspapers. This task was carefully engineered by our corporate communications people: Marshall Post, Henry Von Spreckelsen from the home office and other PR people pulled in from throughout the Bethlehem realm.

A week later, when the press coverage had just about run its course, we staged a two-day open house for everybody else. I think we had over 3,000 people, including steel industry leaders, politicians, leading customers of Bethlehem Steel, educators, and of course our own staff. The receiving and bestowing of invitations became a high-stakes game of office politics.

Only 300 Bethlehem Steel men attended. When I say "only," I mean that we had 150,000 employees, and a great many more than 300 of them had helped to make Burns Harbor a reality. Those who made the trip, aside from top executives, were mostly sales personnel from around the country. They were there to promote the sale of Burns Harbor products to all the customers in attendance.

Tuesday, September 21, was the official opening day. Our guests converged on Chicago from all around the country and were transported to Burns Harbor on a special 19-car train in the early afternoon. In addition, we hired a fleet of buses to bring others in from five major airports from O'Hare to South Bend. We also had parking spaces for 1500 cars.

The guests received a two-hour guided tour of the plant, ending up at the new tin mill warehouse, which was transformed into a festive banquet hall for the occasion. After a huge cocktail party, we served them all a formal lobster and sirloin dinner, followed by musical entertainment by Julie London and Woody Herman and his orchestra. The special train returned its passengers to LaSalle Street Station later that evening, the buses and cars departed, and Stew Cort and

I went to a nearby Holiday Inn in Valparaiso to rest up before doing it all over again the next day!

The following day was "Indiana Day," when we pulled out the stops once again to show off the plant to all the people from Indiana who hadn't been included on the previous list. Heading the Indiana list was Governor Roger Branigin, who was our honored guest. Again, we served a dinner in the tin mill — as big as the night before — and again Woody Herman performed.

I've been to all kinds of industrial and governmental events both in this country and in protocol-conscious cities and palaces abroad. Malcolm Forbes, Aristotle Onassis, Generalissimo Franco — all threw some pretty spectacular parties without batting an eye. But I don't think any of them ever entertained 3,000 people in a steel mill two nights in a row!

In a unique but appropriate way, the greatness of Bethlehem Steel had been displayed to an important national audience. It was a very satisfying moment for those of us who had worked so hard to make Burns Harbor a reality.

Burns Harbor was a reality, but it was far from complete. Over the next several years, subsequent phases of construction would add the necessary facilities for fully-integrated on-site steelmaking: two blast furnaces and three 300-ton capacity basic oxygen furnaces. The combined steelmaking capacity of these facilities was 5.3 million tons of raw steel a year. Other components, such as continuous casters have also been added. All are part of a carefully-mapped master plan to create an orderly and efficient west-to-east, on-line steel plant.

As it turned out, we built those 20-foot coke ovens at Burns Harbor that Floyd Eckhardt and I had studied in Germany. The significance of it was that they were a cheaper way to make coke, since maximum efficiency was an important objective in every facet of the Burns Harbor plant. As usual, I was worried about the doors leaking, but they didn't.

The larger batches of coke could be baked in the ovens just as quickly and then be pushed, yellow-hot, into a car to be quenched in thousands of gallons of water before being screened and fed into the blast furnaces.

The new coke ovens and steelmaking facilities were part of a half-billion-dollar five-year plan for the second phase of construction which we initiated in 1966. In addition, the state of Indiana completed the deepwater development of the Burns Harbor Waterway, which, along with site improvements we made, provided ultra-efficient facilities for receiving raw materials and shipping steel products on Lake Michigan.

In all aspects of the construction of Burns Harbor, the latest high-technology steelmaking enhancements were incorporated. State-of-the-art computer systems assured first-rank efficiency and quality. Anti-pollution features made Burns Harbor a clean operation from the day it opened. Of course, this was a time of rapid advancements in both these areas; because of the naturally slow pace of steel mill construction, we were frequently frustrated and sometimes criticized because it was impossible to keep up with every latest gadget. I certainly disagree with the critics who call Burns Harbor a modern white elephant. After 30 years, it is still at the forefront of worldwide steelmaking, and Bethlehem is doing a good job keeping it there.

1. If I had to select one thing for use as my epitaph, it would have to be Burns Harbor. It was the most challenging accomplishment of my career, and it's still America's newest – and best – steel plant.

1

2. An aerial view of the Burns Harbor plant, looking east along the shore of Lake Michigan. The 160-inch sheared plate mill (right foreground) was the first production unit to go into operation. The 80-inch hot-rolled sheet mill is to the left, and the sheet and tin mill complex is in the background. This photo was taken in September, 1965, when we staged our million-dollar grand opening. Blast furnaces, basic oxygen furnaces, coke ovens and other facilities were built a few years later.

2

Chapter Eleven

Civic Responsibility

The occasion of the fiftieth anniversary of the Bethlehem Chamber of Commerce, on April 10, 1967, enabled me to reflect on an amazing half-century of change in my community, my business and my own career.

In fifty years, Bethlehem Steel had grown from a small but renowned maker of the armor and munitions that had won the First World War to America's 14th largest corporation, with 131,000 employees producing more than 20 million tons of steel in 1967. For 45 of those 50 years, I had witnessed most of that metamorphosis, while working my way up through every fascinating level of the business.

The city of Bethlehem had changed dramatically in fifty years, becoming an integral part of the Lehigh Valley metropolis on the fringe of what was beginning to be known as the Mid-Atlantic "megalopolis," a continuous region encompassing something like one-fifth of the nation's population and business activity. Mass communications, modern highways and other Twentieth-Century inventions were regionalizing America, breaking down provincial barriers. Coincidentally, 1967 also marked the fiftieth anniversary of the "founding" of the modern city of Bethlehem, which had resulted from the merger of Bethlehem and South Bethlehem in 1917, an initial step in that regionalizing process.

I was honored to be the speaker at that Chamber of

Commerce banquet, but not because I was the Chairman of Bethlehem Steel. The honor was shared by my whole family, because the Chamber of Commerce had been founded by my wife's father, Robert Sayre Taylor, a Bethlehem lawyer and banker (he was named after Robert Sayre who for many years had headed the Lehigh Valley Railroad). He and others of his generation, notably Archibald Johnston who was the first mayor of the unified city, had been true visionaries. They set in motion the development of a modern community which became culturally and intellectually cosmopolitan without losing its distinctive heritage.

Bethlehem Steel had of course played a big role in that process. Simply by growing as a company and as an employer, Bethlehem Steel helped turn its hometown into a leading city. In addition, Mr. Grace had taken a great interest in community affairs, and he started a strong tradition of involvement by the company and its people that continues to this day.

I especially stressed community service when I became chairman. We always encouraged civic involvement by employees at every level, and we tried to ensure that the company carried its fair share of responsibility for the community's wellbeing.

As R. S. Taylor's son-in-law (and as Frances Taylor's husband), I became ingrained with a special sensitivity to the pulse of the community. This background served me particularly well during the decade of the Sixties, and, looking back, I'm glad that I was in a position to help address community problems during that very challenging decade.

One of the things that made the Sixties so challenging was the fact that so many people decided to question so many things all in a very short period of time. Virtually overnight, people from all walks of life seemed to become sensitized to society's problems. The switchboard sort of lit up all at once, and the circuits just got jammed. The result was that it became impossible to respond to every concern immediately, and this caused frustration and hostility. In addition, confrontation was kind of a novelty — almost a fad. Everybody, the accusers and the accused, had great stores of unspent emotion. They were battle-ready. Like World War II, a great deal of public energy was unleashed in a very short time period, but, instead of having almost every American on one side of a single cause, they were on every side of a thousand causes. It was a colorful time. And, looking back, it was a very productive time. A lot was accomplished during the confrontational decade of the Sixties.

The concerns of the Sixties that were sweeping America all had special intensity here in Bethlehem. Issues of labor and "big business," ethnic issues, environmental issues — all of these were especially pertinent in Bethlehem. Bethlehem Steel was a target that was easier to hit than the side of a barn. As a result, we soon became very good at absorbing criticism and responding to it in a constructive manner.

One of the big criticisms was that Bethlehem Steel was "dominating" the community. We were seen as Orwell's "Big Brother," somehow invisibly orchestrating the lives of the citizenry. We were seen as Bethlehem's oversized share of the distasteful and distrustful "military-industrial complex," enthusiastically participating in the destruction of the planet.

You've got to have a tough constitution to go to bed each night after listening to or reading about accusations like that!

Those accusations said more about changing times than about changes at Bethlehem Steel. Just 20 years earlier, Bethlehem Steel was one of the best-loved heroes on the World-War-II homefront. After two decades of diligent postwar adaptations, we were suddenly and unexpectedly being characterized as society's arch-villain. It was essential that we find ways to effectively get our story across to the public.

Bethlehem Steel had become the city's largest revenue producer and its most important supplier of jobs. In that respect, we attracted many talented and energetic people who added to the vitality of the community. It could be argued that Bethlehem benefitted simply from having these people in the community, but we encouraged them to be productive, generous contributors to the community as well. My personal philosophy has always been that it's your obligation to serve. I stressed that. We also expanded Bethlehem Steel's corporate philanthropy and community service efforts.

While I was chairman, we tried to encourage everybody to become more civic-minded. We encouraged other companies and the general citizenry to become more involved. A good example was the outstanding success of the drive to fund and build the new public library in 1965. The effort was spearheaded by Roland Adams, the publisher of the *Bethlehem Globe-Times*. He did a great job generating support, and he got widespread citizen and small-business participation.

Another important issue of the Sixties was the advancement of racial equality. The steel industry traditionally has been a patchwork quilt of ethnic groups. Often, after only a generation or two, the sons of immigrant steelworkers could be found in virtually any managerial or professional part of the company. This wasn't happening with black Americans. As with other issues, the sheer size of Bethlehem Steel made us a leading target for criticism.

The problem, unfortunately, had been around for a long time, but the expectation that management had to correct it came about rather suddenly. The civil rights movement during the early Sixties had accomplished a great deal, and all of a sudden the ball was in Ed Martin's court. In 1964, my first year as Chairman, we were challenged by President Lyndon Johnson's Committee on Equal Employment Opportunity to develop and implement a "Plan for Progress."

This document, signed by Johnson and me, enumerated the policies and voluntary efforts we would pursue to broaden equal employment opportunities at Bethlehem Steel. It was a big, new challenge, and yet it was one of many big, new challenges that arose during the decade of the Sixties.

Three years later, in 1967, we were able to report to our employees that significant progress was being achieved. Our work force had a higher black percentage than the general population. And throughout the company, a wide variety of equal opportunity initiatives had become permanent policy. By today's standards, the accomplishments may seem modest — until you stop and remember what things were like throughout our society a quarter-century ago.

As Bethlehem's CEO, I found myself at the front lines of a great many of the community's important causes. The City and local charities got our cooperation as consistently as possible, and I was asked to serve as trustee or director of such leading Bethlehem institutions as Lehigh University and St. Luke's Hospital.

One aspect of Bethlehem's community life which was very important to my wife and me was the preservation of Bethlehem's historic identity. Around the time of that fiftieth anniversary celebration of Bethlehem's municipal consolidation, our reflection on the changes we had seen led us to appreciate that which was in danger of being lost.

My wife Frances was especially concerned about preserving Bethlehem's unique history and heritage. With the exception of our years in Buffalo, she was a lifelong resident and devoted community servant. Her father, R.S. Taylor Sr., in addition to starting the Chamber of Commerce, also founded the Rotary Club and was president for many years of the First National Bank of Bethlehem, which is now Lehigh Valley Bank. Her mother, Caroline Wolle Taylor, had many family ties to the community as well. The Wolles had been members of the Moravian Church all the way back to the founding of Bethlehem. One of them was the stone mason who built the Widows' House which is one of the oldest buildings on Church Street. A cousin of Frances' was J. Fred Wolle, the founder of Bethlehem's famous Bach Choir.

With both sides of her family so intertwined with Bethlehem's history, Frances grew up with a special appreciation for the things that make it such an interesting, attractive and enjoyable place to live. She also grew up with a strong sense of duty to help ensure that the community's special assets would be preserved and enhanced for subsequent generations. These were the things that made me fall in love with Bethlehem and Frances Taylor on the day I arrived here back in 1922. It wasn't the protests and confrontations of the Sixties that got me to throw my considerable corporate weight into the arena of community service; it was my extended honeymoon with one of Bethlehem's fairest and noblest daughters!

As the *Bethlehem Globe-Times* editorialized when she died in November, 1980, "Frances Martin shunned the limelight. However, no top steel executive's wife ever used her power behind the scenes with better judgment. And we suspect that no top executive was ever reminded more about a corporation's responsibility to its home city."

How right they were! First, Frances got the women interested in historic preservation; then, she went to work on the men. Eventually, historic preservation grew to become one of the community's leading civic enterprises, with several strong, well-funded organizations each taking responsibility for preserving and interpreting a specific part of Bethlehem's history.

One of the first historic preservation projects in town was the Goundie House at 501 Main Street. In the mid-Sixties, when Main Street was still a hodge-podge of old buildings with gaudy modern storefronts, word got around that one of Main Street's oldest buildings, the Goundie House (a Federal-style mansion built in 1810), was threatened with demolition. Frances and Christine Sims were the first to wake everybody up to the fact that it had to be saved. They got the movement started. They raised the money to buy and restore it, and this was the beginning of Historic Bethlehem Incorporated.

The movement spread, expanding to the industrial area. One day, Bishop Warneke, a man of vision, and Bill

Johnstone, were standing on the Hill to Hill Bridge looking down on what is now Historic Bethlehem's 18th-Century industrial area in the Monocacy Creek ravine behind the Hotel Bethlehem. Back then, it was literally a junk yard, filled with junked cars.

Bishop Warneke got Johnstone, with his knack for building projects, involved. Then a lot of us got involved. We got the state to give us a million dollars — half a million to restore the Sun Inn on Main Street, and half a million for the Luckenbach Mill in the industrial area. A coincidence is that David Morrison was then working for the Thornburgh Administration in the Department of General Services. They were cutting back on community projects like this, but Dave helped save the Bethlehem project.

The Sun Inn, of course, was a separate organization, headed by Hughetta Bender. A real ball of fire. Her husband, James Bender, had been assistant treasurer and a director of Bethlehem Steel in the Grace era. I was president of HBI for a while. I was followed by Ken Smith, who went on to become Bethlehem's mayor. And I think Dave Dunbar was president after Ken. We got the industrial area cleaned up, and we restored the Tannery and the Waterworks.

HBI has little to do with the old Moravian part of town. The Moravian buildings never needed to be saved because they've been in use and preserved all along. Most of them are 250 years old, and they're still functioning buildings.

Historic Bethlehem has done a great job for almost a quarter century. In many ways, it was ahead of its time. While most older cities and historic places have only recently discovered the benefits of historic preservation, in Bethlehem it's already become a significant aspect of the city's downtown revitalization. Quite a few businesses now thrive because of tourism in Bethlehem, and those of us who live here enjoy it every day. We have a remarkable quality of life for a steel town.

Today, it is clear that the decision to keep Bethlehem Steel headquartered in Bethlehem was a farsighted one. During the decades that we grew to a position of significance in the economic and commercial affairs of the nation and the world, it might have seemed logical to move the company's headquarters to New York. Much of our executive and financial business traditionally has been transacted in New York, even in Charlie Schwab's day. Schwab's primary residence was in New York. Grace spent time a fair amount of time in New York, and so did Homer, who was especially oriented to New York City because it was the center of Bethlehem's shipbuilding operation. When Homer was shipbuilding vice president, up to and during World War II, he was in New York full time. When, I became chairman, I spent an average of about a day a week in New York on business. I had a permanent apartment on the 30th floor of the Waldorf Tower. We might just as easily have moved our whole headquarters to New York, but there were a lot of things about Bethlehem that we liked, so we stayed. When we decided to build Martin Tower, that settled it.

Bethlehem Steel stayed headquartered in Bethlehem, and so did Frances and Ed Martin.

In 1970, the year I retired, Bethlehem was in the early stages of a long-running city revitalization trend that continues to this day. Lots of good citizens have taken great pleasure in helping to nurture this trend, and Frances and I were happy to be amongst them.

1

1. An architect's rendering of Martin Tower. Bethlehem's world headquarters was begun in 1969 and completed in 1972. In addition to centralizing all staff operations in one location, its construction ensured that the corporate operations would stay in Bethlehem.
2. In 1970, I sent Phil and Muriel Berman a 44-inch salmon I caught in the Ristigouche River in Canada. At a dinner party in my honor at their Allentown home, the fish was brought in whole by chef Harry Huffort. Admiring it left to right are Phil, Frances, Muriel and me. (*Phila. Inquirer* photo.)

2

1

1. My wife, Frances Taylor Martin, died in November, 1980. She was under five feet tall and they called her "Mighty Mouse". She was a founder of Historic Bethlehem, Inc. and a trustee of Moravian Academy, among many civic causes. She loved to give a party and was great at entertaining.
2. The historic Goundie House on Main Street in Bethlehem. Frances helped assure its restoration, sparking a preservation movement that has transformed the entire downtown area.

2

1

1. Community leaders. Insurance executive John Heske and NFL football legend Chuck Bednarik, the Bethlehem native who was "the last of the sixty-minute men" with the Philadelphia Eagles, are among the many good citizens I've enjoyed working with for Bethlehem's worthy causes.
2. Leaders at Lehigh. Continuing Bethlehem Steel's long-standing link with Lehigh University, I joined the board of trustees in 1964, and I'm still an honorary member of the board. This group includes vice president for development Paul Franz, me, trustee Ed Curtis, President W. Deming Lewis, trustee Frank Rabold and Provost Albert Zettlemoyer.

2

1

1. Frances and me, with our poodle, Cheri, at home in Saucon Valley.
2. Our 50th wedding anniversary, November 27, 1976. Susan Stout, me, Frances, Barbara Martin Stout, Bettie Martin and Peter Stout.

2

Chapter Twelve

Keeping The Wheels Turning

You've got to do things to keep the wheels turning upstairs.

When I retired at age 68, I was so completely accustomed to being busy all the time, that I stayed right on being as busy and as active as I could.

I still have an office I go to, even though I've been retired more than 20 years. It's on the third floor of the Hotel Bethlehem. When Bethlehem Steel owned the hotel, they gave us offices at the front of the building in Suite 310. Then the hotel was sold, and the new owners wanted our rooms for something else, so they moved us but let us take the room number with us.

Lew Foy, who retired as CEO ten years after I did, rents Suite 310 with me. We share an outer office, and we each have our own inner offices. It's a great place to look after our civic and charitable activities. Foy still has a couple of business ventures he's running, so he puts in more hours than I do.

Shortly after I retired, we organized what came to be known as the 310 Club. It really doesn't have anything to do with Suite 310, except that that's where I handle some of the details. The club's about 20 years old. When I retired in 1970, Art Peterson and Paul Killian and I used to go to the Bethlehem Club every Monday, and after lunch we would roll for quarters. When Morrie Starr retired — he was assistant to the operating vice president — we invited him to come. He was there for about a month when he said, "We could roll for quarters anytime. Why don't we invite somebody to come in here and give us a lecture?"

We all thought that was a pretty good idea, so we started inviting different people. From a beginning with about four or five members, we now have to limit it to 25. All but one or two are ex-officers of Bethlehem Steel. Foy and Donald Trautlein, who followed him as CEO, are members. One outsider is Bill Leckonby, the retired director of athletics at Lehigh; his father-in-law was Arch Johnston, the early steel executive who was the first mayor of the unified Bethlehem, so Leckonby's always been one of the gang. Every Monday, except in summer, we have a private lecture after lunch. Walt Williams was speaker a number of times before he retired and became eligible to join.

It was a few years after I retired that I became president of Historic Bethlehem. That was a big job, and it was nice to have an office in the hotel, because all of the Historic Bethlehem projects were practically right outside the window.

I used to go and see Paul Marcincin about once a week when he was mayor. Sometimes every day. He used to say I was his unofficial local business advisor. I thought he was a good mayor. Ken Smith, who followed him as mayor, is doing a good job, too. Our city's been well run. You can tell just by looking. The South Side, over by the steel plant, is more attractive than ever.

I've been physically active all my life, and I've tried to keep that up in my retirement years.

Like my great-grandfather Edmund Martin II, I kept running. I guess because I didn't have 400 pounds to carry around like he did, I was still running — out here around the roads and fields of Saucon Valley — when I was in my 80's. And I'd swim 50 laps three mornings a week at the YMCA before going into my office at the hotel. Those forms of exercise took discipline; I kept up with golf and fishing because they were my favorite sports and I enjoyed them.

I've mentioned golf enough already, and I'll try not to talk too much about fishing. I was 60 when I took up fly-fishing (although I'd done other kinds of fishing for many years before that.) It occurred to me that I'd need a sport that I could concentrate on when I would no longer be playing golf. I went to Dr. Billy Estes and asked him if he thought I could take up fly-fishing at age 60. He was the chief of medicine at St. Luke's, a fisherman himself, and about 20 years older than me. He told me to go ahead.

Billy Estes's family had a summer place at Pocono Lake, which is surrounded by some of the best trout fishing in the East. The streams flowing into and out of Pocono Lake — the Tunkhannock and the Tobyhanna — are among the best trout streams anywhere. Frances' brother and sister-in-law, Miriam and Bob Taylor Jr., also had a place at Pocono Lake, as did a number of our friends in Bethlehem, Allentown and the business world.

We started renting at Pocono Lake when our grandchildren, Peter and Susan Stout, were first old enough to enjoy the kind of active, outdoor summers I'd enjoyed as a kid. This was around the time we moved back to Bethlehem from Buffalo and no longer had Lake Erie for recreation. I soon became interested in learning to fly-fish the outstanding trout streams around Pocono Lake.

My fishing mentor was Arthur Frick, another Bethlehem friend who'd been at Pocono Lake for a number of years. Artie Frick was around Billy Estes' age, and he too assured me that I was not too old to learn to fish, being twenty-some years younger than they were. Artie had been president of the Allentown-Bethlehem Gas Company, which later merged with UGI. Between Artie and his son-in-law, Jim Cox, I received just about the finest instruction in fly-fishing available.

In 1964, Frances and I bought our own place at Pocono Lake. Pocono Lake was built in 1899 by the Tobyhanna Water Storage and Supply Company to make ice in the days before artificial refrigeration. The venture was started by William J. Daub of Easton, whose grandson, Bill Daub II, was a contemporary of mine, and whose great-grandson, John Daub, operates the Brown-Daub automobile agencies in Easton. He still has his great-grandfather's house at Pocono Lake, up at the end of the lake near where the ice houses used to be. The ice business apparently continued until about 1940, but in the meantime the lake became a summer resort. Today, about 130 families have camps around the shores of the lake. Most of them are still very rustic.

Over the years, a number of Bethlehem Steel colleagues have been neighbors at Pocono Lake. Benjamin Etter, who was head of the credit department, was there in the Forties and Fifties. John Long, a noted historian and Bethlehem's long-time manager of publications, was there for many years. Don Barnum, who was assistant vice president of industrial relations, was two doors down from Etters. Tom Church, retired vice president of transportation, is there. And Jim Cox, retired superintendent of forging, has been there since he married Ruth Frick. Ruth's brother, Jack Frick, had been in the executive office at Lackawanna. Today, Pocono Lake has an annual sailing regatta in Jack's memory.

Our camp was built by the Kleins, who were among the earliest summer settlers, and we bought it from the Wright family. Like most of the places at Pocono, it was old and dark and had low ceilings and doors. It also had a magnificent old stone fireplace which supported the entire house. We went to great lengths to restore the camp so that we would have the best of both worlds: the original fireplace and traditional architectural lines along with a brighter, more modern interior with a better layout and more headroom. I was always very pleased that we accomplished that without demolishing the camp.

The first time Artie and Ruth Frick came over to visit us at our new camp they brought a dozen cocktail glasses. "The glasses in this camp are too small!" Artie said, referring to some previous visit. You usually got the contents of the camp when you bought it, and most Pocono camps were notoriously ill-equipped. Most still are. We like to keep things simple there. I'm sure that's why I enjoyed it so much there; during the years that I was chairman of Bethlehem Steel, about the only place I could escape the formalities and trappings of my position was at Pocono.

Our friendship with the Fricks prompted one additional improvement at our camp. When we bought it, it shared a common driveway with two neighboring camps. You had to do a lot of backing up. The Fricks teased me about that. They said, "Neither of us can back out of here!" So I put in my own driveway with a turn-around in front of the house, and we named it "the Ruth Frick Circle."

Artie Frick continued to fish into his 90s, and I would've too if I hadn't had a stroke at age 88. But for years, I used to catch trout until I couldn't give them away any more. I caught 80 or 90 a year.

I still go to Pocono for a short visit every summer, but after Frances died, in November 1980, I turned the camp over to my daughter Barbara. I often said it was the smartest thing I ever did. She's very involved in community activities there and was chairman of the religion committee for many summers. There's still not a summer that goes by that she doesn't invite up at least one Moravian minister as guest preacher.

I fished many places besides Pocono. One of the first trout-fishing experiences I had was with Bill Price, the CEO of Westinghouse. He took me to the Rolling Rock Club out near Pittsburgh and sent me out with a guide to fish. I got my lures caught in the trees a lot. I was up in a tree more than I was casting. He helped me learn how to avoid the trees.

Usually on a private trout stream you have to mark down

how many fish you caught so the fish committee or the warden can keep track of the fish population. At Rolling Rock, at each pool you were in you had to record your catch. If the pool was fished out, it would be stocked overnight and full of fish again the next day. I went to Rolling Rock several times and enjoyed it very much.

Bill Price also introduced me to salmon fishing in Canada. Westinghouse had a lot of business in Canada, so he was familiar with all the best fishing spots. We fished the Upper Jupiter River, which is one of the best places for salmon fishing anywhere. In those days, you got there by travelling upstream in a canoe pulled by horses, since they still had no roads. We usually went during the second week in June and stayed for about five days. Sometimes, I went again in August and took Frances with me.

Another exclusive fishing enclave was the Ristigouche Club, which is also up in Canada. I got involved there through Bill Price too. He invited me to the club's annual meeting in New York City and I put in an application and joined. The fees ran about $8,000 a year whether you fished or not. In addition, each year when I went up I hired a guide. He was one of the best guides up there, and I used to give him a $600 tip.

Frances always went up to the Ristigouche with me. After our automobile accident, she couldn't sit in a canoe anymore, so we stopped going.

More recently, my fishing companions were George and Priscilla Payne Hurd. They both enjoy salmon fishing, and we took some memorable trips together while I was still able. Priscilla also serves on the St. Luke's board with me.

In 1989, from the after-effects of my last stroke, I slipped and fell on the front step of my house in Saucon Valley. I was answering the door, and somehow I just slipped. Since then, I've had to use a walker to get around, but I can go anywhere the walker can go. I have two walkers: one is the kind you pick up and put down; the other is on wheels and has a handle like a bicycle. It even has hand brakes.

Fortunately, my current house has an elevator. We moved there in the late Seventies. It was too difficult for our daughter Bettie to climb the stairs at Oak Knoll when she came to visit, so when this house became available we took it. It's the only house on our private road that has an elevator.

There are eight houses on our private road. They were originally owned by the steel company for retired executives to rent. I built them. They're right next to Weyhill. But after I moved in the steel company made us buy our houses. Now we have an association to run the place, plow the snow and so forth. Tom Stott, who lives a few doors away heads it up. He married Mr. Cort's daughter, and like his father-in-law became general manager at Sparrows Point and then vice president for steel operations. We've joked that I always wanted my out-of-town assignment to be at Sparrows Point because of its reputation under Mr. Cort, while Tom always wanted to go to Lackawanna because of its reputation under me. He was a top-notch steelmaker and a good manager.

Ever since that last fall, I've had to have someone living with me all the time. Before that, I had one of these alarm systems around my neck; you push the button and the cops show up in five minutes. Since 1991, my live-in helper has been Debbi Stettler. She runs the house and helps me with everything, and she takes an interest in all my activities. We discuss everything. It's great, because I'm still living on my own, in my own house, but I'm not alone.

Debbi does all my driving, now. Sometimes, when I go to banquets and things, she'll come right along as my dinner date. We get along great. I shout at her, and she shouts right back!

Of course things go smoothly at home also, because of the devoted service of my faithful housekeeper, Lil Schrantz. She's been with me so long, she was originally trained in our other house by my wife Frances. That was about thirteen years ago.

At the office, my secretary Erma Buck keeps things straight and tied together. She helps me organize my schedule, and she takes care of most of my tax and financial matters. I still get asked to become involved in a great variety of charitable and community causes, so there's always plenty of work for Erma and me both.

Over the past couple of years I've received a number of very nice honors and awards. It's nice to know, as I hit age 90, that people out there remember what I did. The honors for my community service have been especially meaningful. Historic Bethlehem, the Chamber of Commerce, the Society of Fundraising Executives and other groups have all held dinners in my honor within the last year or so.

It makes me feel good about Bethlehem. Bethlehem gave me all the things that made my life a happy one: my career, my family and my friends. I'm glad I've been able to give something back.

My only regret is that my wife, my parents, my father-in-law — and all the others who helped mold Ed Martin during my 90 years in life's loop — can't be here to share and enjoy each award with me.

1

2

3

1. Keeping the wheels turning. In my office at the Hotel Bethlehem in August 1992, I made a videotape of remarks to be screened at the annual Ed Martin Golf Tournament awards banquet at Briarwood Country Club, the club I built on the shore of Lake Erie. Bill Gignac of Bethlehem Steel's public affairs office conducted the interview.
2. Lew Foy keeps the wheels turning, too. Bethlehem's seventh CEO now shares the suite 310 office complex with me at the Hotel.
3. Still making speeches. In March, 1992, as part of Bethlehem's 250th anniversary celebration, I was honored by the Bethlehem Chamber of Commerce with a standing ovation for my 70 years of community service. I received the Frank Marcon award.